Elijah Whitney

Asiatic Cholera

Elijah Whitney

Asiatic Cholera

ISBN/EAN: 9783337315528

Printed in Europe, USA, Canada, Australia, Japan

Cover: Foto ©berggeist007 / pixelio.de

More available books at **www.hansebooks.com**

ASIATIC CHOLERA

A TREATISE

ON ITS

ORIGIN, PATHOLOGY, TREATMENT, AND CURE.

BY

E. WHITNEY, M. D.,

AND

A. B. WHITNEY, A. M., M. D.,

LATE PHYSICIAN AND SURGEON
To Diseases of Women in the North-Western Dispensary, Visiting Physician, Etc.

NEW YORK:
M. W. DODD, PUBLISHER,
No. 506 Broadway.
1866.

Entered according to Act of Congress, in the year 1866, by
A. B. WHITNEY, A. M., M. D.,
In the Clerk's Office of the District Court of the United States, for the Southern District of New York.

E. O. JENKINS, STEREOTYPER AND PRINTER,
20 NORTH WILLIAM ST., N. Y.

DEDICATION.

TO PROFESSORS POST, VAN BUREN, METCALF, AND BEDFORD.

FOR those lucid Clinic illustrations and faithful instructions during a three-years' course in the New York Medical University, particularly the critical Pathological knowledge there inculcated, and consequent professional success, the youthful author is indebted.

Knowing they will agree with him, that his appreciation of their valued services, and his gratitude for the same, can be best acknowledged in his attempt to benefit suffering humanity, he would here publicly acknowledge the pleasure and benefit received from their instructions during his College course, and beg their acceptance of his sincerest gratitude and affection.

TO THESE ABLE INSTRUCTORS THIS VOLUME IS CORDIALLY DEDICATED BY THE JUNIOR AUTHOR.

A. B. WHITNEY, M. D.

PREFACE.

The following pages are the result of investigations and the collection of facts and arguments from a great variety of sources, originally made and presented in aid of the discussions on the subject during the past six or eight months.

The most eminent and reliable authorities for nearly half a century, that is, from 1832 to 1865, including the late reports from India, have been carefully examined, and such late discoveries, facts, and arguments collected, as seemed to throw light upon the subject, or in any degree to indicate or direct to a general principle of practice.

The various experiments instituted for the cure of the disease have been carefully investigated, and the principle evolved explained whenever any advantage was derived from the same.

In all these we have diligently searched for the cause of the failure of "remedial agents," so uniformly admitted, and have endeavored to present the results clearly and fully in the body of the work.

Our statistics are collected from reliable sources,

PREFACE.

The following pages are the result of investigations and the collection of facts and arguments from a great variety of sources, originally made and presented in aid of the discussions on the subject during the past six or eight months.

The most eminent and reliable authorities for nearly half a century, that is, from 1832 to 1865, including the late reports from India, have been carefully examined, and such late discoveries, facts, and arguments collected, as seemed to throw light upon the subject, or in any degree to indicate or direct to a general principle of practice.

The various experiments instituted for the cure of the disease have been carefully investigated, and the principle evolved explained whenever any advantage was derived from the same.

In all these we have diligently searched for the cause of the failure of "remedial agents," so uniformly admitted, and have endeavored to present the results clearly and fully in the body of the work.

Our statistics are collected from reliable sources,

are very brief, and introduced in aid of the main object,—the establishment of a general principle of practice.

The different modes of practice are from the most distinguished authors of the different Schools of Medicine, and non-professional gentlemen; condensed and exhibited mainly in their own language, to show their conformity or non-conformity to the Pathology of the disease.

In all we have kept constantly in view the pathology of the disease, whose "dictates" have governed us in the exhibition and establishment of a general principle of rational practice, confirmed by observation and experience, which, if accepted and carried out by the profession, we hope and trust will save a very large proportion of those attacked by "this most acute of acute diseases."

AUTHORS.

CONTENTS.

CHAPTER I.

Sec. I.	Origin and Development............	7
Sec. II.	Progress and Fatality.	20
Sec. III.	Causes—Propagation.	34

CHAPTER II.

Sec. I.	Pathology........	45
Sec. II.	Phenomena, or Symptoms....	55

CHAPTER III.

Sec. I.	Unsuccessful Modes of Treatment. — Venous Transfusion Explained......	64
Sec. II.	Physiological Condition of the Blood, Its Non-Aeration, or Non-Oxydation. Maxims of Rational Practice Suggested.	91
Sec. III.	Different Modes of Treatment.........	130
Sec. IV.	Statistics. Percentage of Loss. Variable Results—their Cause.....	166

CHAPTER IV.

Sec. I. GENERAL PRINCIPLE OF RATIONAL PRACTICE, DICTATED BY THE PATHOLOGY OF THE DISEASE, CONFIRMED BY OBSERVATION AND EXPERIENCE.................. 178
Sec. II. REMEDIES, RECIPES, ETC.................. 188
Sec. III. PROPHYLAXIS, OR MEANS OF PREVENTION.. 203
Sec. IV. FORMULÆ—PREPARATIONS, ETC............ 213

ASIATIC CHOLERA.

CHAPTER I.

Section I.—Origin and Development.

Epidemics have occasionally prevailed in all ages. Sometimes they have been circumscribed in their influence, and limited to particular localities; while at other periods they have taken a wider range and extended over larger sections, inflicting the most lamentable results, and augmenting the bills of mortality to an incredible degree.

The earlier writers have given some account of these diseases, which have occasionally prevailed as very fatal and devastating epidemics; surpassing all other diseases in their mysterious origin, in their rapid extension, and in the duration of their prevalence. In the East,—in Egypt, and on the eastern border of the Mediterranean, fearful epidemics have prevailed from time immemorial. They have often proved very destructive, especially in the Middle Ages, and as late as the sixteenth and seventeenth centuries. During the prevalence of the "Pestis," which raged throughout Europe between the years 1347 and 1350, according to computation, a fourth part of the inhabitants of this part of the globe was carried off. The estimates of the vast numbers swept away by its repeated occurrence and prevalence appear quite incredible.

During the time it raged at Marseilles in 1720, it is reported that in the Charity Hospital there were admitted from October 3d to February 28th, 1,013 patients, of whom 585 died; and during the same period, in another hospital, there were admitted from October to July 3d, 1,512 patients, of whom 820 died. The population of Marseilles previous to the occurrence of the disease was estimated at about 90,000, of whom 40,000 died; leaving only about 10,000 of the whole population who had not been attacked or in any way affected; so that the record shows the appalling mortality of fifty per cent. of those who were attacked.

The bills of mortality in 1770 and 1771 were as appalling as any arising from epidemics of a later day. A very extended notice of the "Pestis" as it raged in Moscow in the year 1771 is given by M. Gerardin, who, quoting from the published statistics, observes: "In April, the deaths were 744; May, 851; June, 1,099; July, 1,708; August, 7,268; September, 21,401; October, 17,561; November, 5,235; December, 805; making a total in nine months of 56,672, which is considerably less than the estimate given by De Mertens, who thinks the whole number carried off by this pestilence, from the city alone, cannot be less than 80,000. These statistics bear a striking resemblance to those of the Epidemic Cholera, whose fatality is materially varied by the seasons of the year; the greatest being usually at the close of Summer or the beginning of Autumn. There are, in short, many points of resemblance in this and former epidemics to that of the Cholera, which naturally lead to the supposition that all have had a common origin, if, indeed, they be in many respects dissimilar.

Their pestilential character, their extended influence, and their great fatality, rendered their appearance and progress a special terror to physicians, and melancholy apprehension to the people. They seem to have been regarded as the manifestation of an invisible power, which directed and guided " the pestilence that walketh in darkness" and "the destruction that wasteth at noon-day;" a visitation or chastisement over which human ingenuity and medical skill had little control. Under these impressions, the earlier physicians labored and endeavored to satisfy the great mass of mind that these occasional and special developments of disease arose from natural causes, and were subject to certain natural laws. They ascribed their origin to the commingling of some specific poison in the food, and drink, and air, which, through these "media," was received into the system.

Subsequently, they seem to have made some advance on this theory, and considered the extreme Summer heat—especially the intense heat of the sun in a dry season—the emanations from stagnant waters, and the miasm exhaled from the soil, and from putrid bodies of animals, as the chief causes of all epidemics. These views prevailed for a very long period, and have undergone no very remarkable change from the observations and discoveries of centuries.

Modern and quite recent writers have advanced nearly the same doctrines, embracing, however, the principal sources of insalubrity — the malarious and miasmatic influences; and have assigned as the cause of epidemics, especially that of Cholera, a peculiar constitution of the atmosphere, and certain predisposing causes combining with each other, so that an association or union of these two independent and individual causes are necessary and essential to the production of the disease.

Eminent scholars and pathologists have, during the century last past, patiently searched for its final cause, without arriving at any better, wiser, or more satisfactory conclusion than the earlier writers, who regarded it a poison, commingled with the food they ate, the water they drank, and the air they breathed. The modern writers, according to the more popular views, almost universally adopt the hypothesis that the remote or final cause of the Cholera is a specific poison; for at no period has a person in good health in this or any other country been known in a few minutes to be shriveled up, his face and extremities to turn purple, his whole body to become of an icy coldness, and with or without vomiting a peculiar fluid, like rice-water, to die in a few hours, except under the influence of poison. That this disease, so appalling and destructive in its effects, and so mysterious in its wanderings, should spread over countries in respect to climate, soil, geological formations, and as to the moral and physical habits of the population, so utterly opposite to those where it first originated, is only explicable on the hypothesis of its propagation on the principle of a specific disease—poison.

How and in what manner it travels has not been satisfactorily determined. Whether independent of any and all human agency, or absolutely dependent on ordinary communication and intercourse of tribes, and peoples, and nations, is as yet unsettled. It is, however, a matter not of so much consequence as the fact that, in all its nomadic life, it retains unchanged its youthful disposition, vigor and energy. It seldom shows any inclination to associate, or coalesce, or even adopt the milder habits and manners of others.

Perhaps some idea of its character may be obtained from a microscopic view of its birthplace and its surroundings. Whether the locality of its irruption in 1629, or that of 1817, whence it spread over the greater part of the globe, be entitled to the unenviable distinction of fostering its gestation, concealing and protecting its birth, and nursing its infancy, is immaterial;—since the similarity of these localities strikingly illustrates its cause and ultimate development.

On the north side of the island of Java, about 6° S. lat. and 107° E. long., near the mouth of the river Jacatra, is situated Batavia, in the midst of swamps and marshes, surrounded by trees and jungle, which prevent the exhalations from being carried off by a free circulation of the air, and render the town peculiarly obnoxious to marsh miasmata. Besides this, all the principal streets are traversed by canals, planted on each side with rows of trees, over which there are bridges at the end of almost every street. These canals are the common receptacles for all the filth of the town. In the dry season their stagnant and diminished waters emit a most intolerable stench, while in the wet season they overflow their banks and leave a quantity of offensive slime. From these united causes, it is not surprising that Batavia has been considered the most unhealthy spot in the world, and has been designated the store-house of disease. According to Raynal, the number of sailors and soldiers alone who died in the hospitals averaged 1,400 annually for sixty years, and the total amount of deaths in twenty-two years exceeded a million of souls. The city was inclosed by a wall of coral rock, with a stream of water on each side within and without. Few Europeans, however, sleep within the town, as the night air is considered very baneful. The inhabitants, possibly, as an antidote against the noxious effluvia arising from the swamps and canals, continually burn aromatic woods and resins, and scatter about a profusion of odoriferous flowers, of which there are great abundance and variety. During the prosperity of the Dutch East India Company, Batavia obtained the title of Queen of the East, as the resources of all other districts were sacrificed to its exclusive commerce. Here, in this noted locality, was the Cholera bred and reared in 1629, under circumstances of great significance, admirably adapted to convey some idea of its cause and character.

A learned professor, speaking of the diseases of India, observes : " CHOLERA is the most acute of acute diseases. It seems to have existed in Batavia as far back as 1629 ; and it has been known to prevail as an occasional epidemic in India at different years and places from 1774 to 1817. Since then it has been endemic, and is a disease whose germs are essentially maintained in, or upon the soil. It annually recurs at many of our large stations, commencing generally at the beginning of the hot season, but sometimes occurring in the rainy and cold season. Its greatest proclivity to propagation is amongst populations living in low, damp, crowded, and ill-ventilated situations, especially if the water supply is impure. Nearly all the diseases fatal in India are accompanied by profuse discharges, with which the air, water, linen, bedding, closets, walls of hospitals, and barracks become more or less infected; so that the 'Materies Morbi' come into contact with all the inmates of buildings where the disease prevails."

Its origin, or reappearance in 1817, is not in any respect essentially different from its earlier development on the Jacatra. The River Ganges, in India, like the Nile in Egypt, flows for a long distance through a low, level country, which it annually inundates. Dividing its waters about 200 miles from the sea, the Delta of the Ganges commences and continues its variegated and checkered surface, till, approaching the borders of the sea, it presents a peculiar aspect, being composed of a labyrinth of creeks and rivers, called "The Sunderbunds," with numerous islands, covered with the profuse and rank vegetation called "jungle," affording haunts to numerous tigers and other beasts of prey. This large river, "a Deity of the Hindoo," is subject to an annual freshet, often rising to the height of 32 feet in the month of July; when all the lower parts of the country adjoining the Ganges, as well as the Burrumpooter, are overflowed for a width of one hundred miles; nothing appearing but villages, trees, and sites of some places that have been deserted. Here in this vast pest-house, where every conceivable vegetable and animal substance is left upon the soil by the retiring inundation, exposed to the heat and dews of a tropical climate—where, too, noisome and infectious diseases have prevailed for centuries, the Epidemic Cholera is said to have arisen and acquired its strength and full development. A fit origin for a fatal and devastating pestilence.

To this low, insalubrious, and festering locality, this vast pest-house, where so many noxious and noisome diseases are generated, and where so many epidemics have arisen and so often swept over the surrounding regions with most fatal and desolating effects, is ascribed the birthplace of the Epidemic Cholera of 1817. Here it is said to have first made its appearance at Jessore—a populous town in the centre of the Delta of the Ganges; whence attaining its growth and power, it has extended its influence as from a common centre, and marked its progress with hecatombs of victims in the direction of almost every point of the compass.

Here we may remark, that it is not our intention to travel over the whole ground embraced by the subject under consideration; but, on the contrary, to present in this treatise only a cursory view of a few prominent features which may interest and aid in the important object of deducing from the pathology and the varied phenomena of the Cholera some general principle of practice. For this, and to this, our labor and our investigations are directed. Availing ourselves of every source of information within our reach, and relying in part on the observations and experience of others, we shall aim to present such facts and arguments as will shed light upon the subject, and aid in the accomplishment of this desirable object. However difficult this may appear, it is nevertheless believed to be within the province of science and unbiased reason.

Section II.—Progress and Fatality.

The disease in 1817 appeared on the Delta of the Ganges, and gradually extending its influence, swept over various countries with terrible severity. Having here acquired its full development, and manifesting an indomitable determination to itinerate, it starts upon its lethean errand, and soon shows a capacity and power to overcome every obstacle opposed to its progress, and to pursue its course unchecked and even unretarded by any natural or artificial barrier. It soon traversed India, and in the succeeding season spread over adjacent countries, visiting in 1818 the Indian Peninsula, the Burmese Empire, the Kingdom of Aracan, and the Peninsula of Malacca. In 1819 it reached Sumatra, Singapore, and various other islands situated along the coast on either border of this vast peninsula.

During the year 1820, pursuing steadily its progress eastward, it reached Tonquin, Southern China, Canton, the Philippine, and numerous other places and islands in that direction. In 1821 it visited Java—the place of its earlier nativity—Madura, Borneo, and many other places in the Indian Archipelago. During the years 1822, 1823 and 1824, it continued to spread over the vast and populous regions of Central and Northern China and the numerous islands upon the coast, and in 1827 prevailed in Chinese Tartary, leaving few places in all these different countries on the continent, or even on the islands bordering on the eastern coast, unscathed by its terrible ravages and depopulating influence.

During the same period, its progress westward has been uninterrupted, and attended with results no less remarkable. It has baffled all attempts to check or even retard its onward course, or mitigate its appalling effects. In July, 1821, it had reached Muscat in Arabia, and thence extended its influence to the populous cities and villages along the Persian Gulf. During the same season it appeared in Persia, and continued to ravage the principal cities and towns of that empire for four successive years. At Bassorah and Bagdad it broke out in July, 1821, and thence extended its desolating influence westward to the Red and Mediterranean Seas, carrying off vast numbers of the inhabitants of the populous cities of Mesopotamia, Syria, and Judea.

In 1822 it prevailed among the nomadic and Tartar tribes in Central Asia and in the northern Persian Provinces, and in 1823 broke out on the Georgian frontiers of Russia, at Orenburg on the River Ural, and at Astrachan on the Volga. Here its western course was apparently interrupted. There was, for a short period. an interval of complete immunity from its presence. Along the border of the Russian Provinces the disease had entirely disappeared, and seemed inclined to retrace its course and return to the home of its birth. But the fond anticipations of Europeans were disappointed; the destroyer was not to be arrested and turned back in his progress over the earth; his march was onward, his demands imperative.

Hence, in the month of June, 1830, the disease reappeared in a Persian province on the southern shore of the Caspian, and again at Astrachan, on the Volga, in July, where it prevailed with such unwonted violence that, before the close of August, more than 4,000 persons had died of it in the city, and 21,270 in the province. From its interval of repose, it would seem to have recuperated its strength and vigor for the lethean work awaiting its progress. Ascending the Volga, it reached Moscow, became prevalent there in September, and continued with great severity till February, 1831. Here it attacked, in the city, about 9,000 persons, of whom more than one-half died. Continuing its advance, it reached Riga about the middle of May, and St. Petersburg on the 26th June.

From Astrachan it also directed its course towards the northern coast of the Black Sea, and thence along the course of the rivers into the central parts of Russia. It reached Poland in January, 1831, accompanied the Russian army in its various marches and encampments during the subjugation of that country, and proved very destructive in Warsaw and many other places during April and May. It appeared at Dantzic in May, and in June at Lemburg, Cracow, and various other places and sections of country, extending through Gallicia, Hungary, and reaching Berlin and Hamburg in August and September, and Vienna about the same time.

Smyrna was visited in September, and Constantinople soon afterwards. It is reported that the pestilence was conveyed by a caravan from Mecca to Cairo in August, 1831, some thousands having died on the road; and, by the middle of September, 10,400 Mohammedans, besides Jews and Christians, had died of it in this latter city.

Passing from the western coast of the continent, on nearly the same parallels of latitude, it found its way over the Northern Sea to the British Isles, and made a lodgment, first, on the northeastern coast of England, in October, 1831, at Sunderland, situated in latitude 55° north, whence it prevailed and extended its influence over this section, evincing the same malignant and lethean character it had manifested in its progress over the continent. Its course thus far has been marked with unparalleled fatality.

It made its first appearance in Scotland, at Haddington, in December, 1831, and at Edinburgh in January. In these and various other places it prevailed for some months, and, as warm weather came on, increased in severity, and carried off a large percentage of those attacked. After spreading thus over the northern section, and rioting for months in the more populous cities and towns, it made its appearance in London on the 14th February, 1832, where it found an abundance of material for recuperating its strength and multiplying its forces, and soon after spread over various other places in the United Kingdom, inflicting the most appalling bills of mortality. In short, its progress over this country has been attended with the same destructive influence and the same lamentable consequences as on the continent. No change, modification, or softening of its disposition or character has arisen from its passage over the Northern Sea, nor from the refreshing influences of a purer atmosphere.

It appeared in Calais on the 12th, and at Paris on the 26th of March, 1832, where it continued in these and other cities and villages for some months with its accustomed severity. During the season it raged throughout the vast empire, and swept away an immense number of its inhabitants. During the succeeding years, 1833 and 1834, it traversed Spain, and proved very destructive in many of its larger cities and villages.

In the mean time, continuing its course from the British Isles westward, unchecked by the prevailing western winds and the broad expanse of the Atlantic Ocean, over which it passes a distance of nearly three thousand miles, and makes its first appearance on the American continent at Quebec, Lower Canada, on the 8th June, 1832, and reaches Montreal on the 10th of the same

PROGRESS AND FATALITY. 27

month. From these cities it rapidly spread in all directions, prevailing in the towns and villages on the St. Lawrence and its tributaries, and soon extended along the chain of lakes, dividing the Provinces from the United States, visiting the principal ports on either shore. It exhibited in all these places its peculiar epidemic character, and proved excessively violent and fatal wherever it appeared.

Its first irruption in New York was on the 24th June, 1832, sixteen days after its appearance at Quebec, and at Albany, midway between the two former cities, on the 3d July. From New York it extended its influence to Flatbush and Gravesend, Long Island, where it appeared on the 5th July, and on the same day and date at the city of Philadelphia. It broke out at Rochester on the 12th and at Buffalo — July.

Thus, while it was making its way westward along the great chain of lakes, towards the arteries of the Great West, it was, at the same time, steadily pursuing its uninterrupted course along the coast, visiting the

3*

main cities, and spreading from these as from common centres over the intermediate towns and villages. In its progress it reached Baltimore on the 22d August, and the City of Washington on the 28th of the same month.

Thence it continued its course to Richmond, Norfolk, Edenton, and various other cities along the Atlantic and Gulf coast.

It appeared at New Orleans in the Autumn of 1832, during the existence of a severe epidemic of yellow fever, and apparently subsided on the disappearance of the fever. Sporadic cases, however, occurred during the Winter, and in the opening of Spring it broke out with unwonted vigor and severity, and thence spread, according to its accustomed laws of itineracy, along the rivers into the interior of the States bordering upon the Mississippi and the Gulf coast, and raged throughout Louisiana and Texas with unusual violence and fatality.

In 1832, 1833, and 1834 it prevailed throughout the Mississippi Valley with great fatality, especially in the principal cities, villages and towns situated upon its navigable waters. Here, after intervals of entire immunity from its presence, it occasionally reappeared in some of the larger cities with renewed vigor and power, and swept off vast numbers of the inhabitants. In no section of the States have greater numbers, compared with the whole population, fallen victims to it than in the fertile and sparsely settled prairies of the South and West.

Thus, from the North, and at a later date from the South, extending its influence along the principal rivers into the interior, it swept over the States, prevailing in some places in the Valley of the Mississippi as late as 1836. In short, it reappeared in 1834 in many cities and places where it had before prevailed, and again spread over a considerable portion of the country with unprecedented fatality.

In 1833, the disease appeared at Havana and Matanzas, and prevailed on the island for several months with great fatality, especially among the colored people. During the same season it appeared in August at Tampico, Campeachy, Vera Cruz, and the city of Mexico, proving especially violent and destructive in these and other cities of the Republic. In Central America it is said to have attacked the army, and in a very short period to have swept away a very large proportion of its officers and men.

Thus, it appears that the epidemic or Asiatic Cholera, from its first irruption on the northern coast, spread over the greater part of the North American Continent in the space of two years, and has several times reappeared in different sections in its peculiar malignant character, spreading on each occasion over a greater or less extent of territory with the same uniform and destructive influence. Neither time, nor science, nor professional skill has thus far appeared to soften its character, or mitigate its severity.

When the disease had fully assumed its epidemic or malignant type in India, in 1817, its rate of mortality was everywhere in that vast territory excessively high.

PROGRESS AND FATALITY. 31

According to the most reliable reports, the cases occurring in the earlier period of an irruption were generally fatal, few only surviving the attack; while of those occurring when the disease was on the decline, a greater proportion recovered. We read of numerous instances where one-third, one-half, two-thirds, and even nine-tenths of those seized with Cholera perished, and again of some places where one-fifth, one-fourth, and in some instances one-third of entire populations were cut off in a very short period by this disease. But without attempting to give the statistics of cholera in this part of the world, or even in Europe or America, we may present a few instances of mortality, going to show the great percentage of loss by this singular disease during its ravages from 1817 to 1837.

In Siam, it is said 20,000 persons fell victims to it in twelve days. The inhabitants are remarkable for their uncleanly habits, and crowded, ill-ventilated tenements.

In Sicily, 16,000 died of cholera in 1832,

at Cataria; in Palermo, 40,000. These cities are represented as being filthy in the extreme, and the personal habits of the people so uncleanly, and the houses so crowded, that it is a matter of surprise the mortality was not greater.

In Bassorah and Bagdad, situate in low, unhealthy localities, and exposed to a damp, insalubrious atmosphere, which, in the warmer season, is often essentially impregnated with miasmata and offensive exhalations from animal and vegetable decomposition, both within and without their inclosures, it is affirmed that more than one-third of their entire populations were carried off in less than one month.

In the Province of Caucassus, out of 16,000 attacked by the disease, 10,000 died. In Russia, out of 54,000 attacked in 1830, it is said more than 31,000 died.

In Hungary, it is reported that the whole number affected by the disease was about 400,000, of whom more than one-half died.

It is officially stated that the total number—the military excepted—of those affected with cholera in France, from its first appearance at Calais, March 15, 1832, to January 1st, 1833, is 230,000, and the deaths 95,000.

In England, the whole number of cases of Cholera is reported to be 49,594, and the number of deaths 14,807. In London there were 11,020 cases, of which 5,274 were fatal. In Wales there were 1,436 cases, of which 498 proved fatal. In Ireland, from its first irruption in 1832 to March, 1833, there had occurred 54,552 cases of cholera, of which 21,171 were fatal.

In Quebec, from June 9th to September 2d, 1832, there had occurred in that city alone no less than 5,783 cases of cholera, of which 2,218 were fatal. In Montreal, from June 10th to September 21st, there were 4,440 cases, and 1,904 deaths reported.

In New York, from July 4th to August 28th, in 1832, there had occurred 5,814 cases of cholera, and 2,935 deaths by the same disease. In Philadelphia, from July 4th to August 28th, 1832, there were reported 2,314 cases of cholera, of which 935 were fatal.

In many of our Southern and Western cities and villages the percentage of loss from the prevalence of cholera is considerably higher than the general average, compared with the data given above. The mortality varies materially in different localities, and, indeed, becomes very much augmented by the prevalence of those influences which particularly favor the vegetation, and are especially concerned in the production of zymotic diseases, whether in the lower or higher latitudes.

SECTION III.—CAUSES—PROPAGATION.

THE remote or final cause is essentially of miasmatic origin, developed under certain atmospheric and terrestrial local conditions, not well defined or fully understood. In its nature and essence, it constitutes a peculiar disease-poison, which is now generally admitted to be, in one way or another, absorbed, and infects the blood, inducing a primary disease of this vital fluid, and directly depressing and deranging the ganglionic system of nerves. To its general character, and the circumstances under which it is generated and in which it operates in producing the disease, we have alluded in speaking of its origin.

The predisposing causes are as numerous as the varied influences which operate to depress the general health. The insalubrity of the atmosphere may be regarded as a general, and, perhaps, the most extensive predisposing cause. In this state, its vital element becomes diminished or impaired to such an extent as to render it incapable of sustaining the normal and healthy functions of the system in their most vigorous condition. Hence, the foul and noisome air of close, ill-ventilated apartments becomes very depressing and baneful; a direct and effective element, often, in constant operation in generating and producing the cholera, typhoid fever, or other deadly maladies. This is not unfrequently the case on board some of our emigrant ships, when hundreds of human beings are stowed away between decks without the means of efficient ventilation, disinfection, or other mode of expelling the noxious principle. Though the germ of disease may be ever on board, it does not vegetate and come forth and rapidly acquire its activity, vigor and power, unless the localizing influence vivify, foster and nurture its development. This is fully confirmed by the recent arrival of two steamers with cholera on board.

The *England*, and a few days later the *Virginia*, with crews and passengers all in perfect health, departing from a healthy port where no cases of cholera were known to exist, and after being out at sea six or eight days under the influence of a cool, invigorating atmosphere, were surprised by the sudden irruption of cholera on board. It breaks out among the steerage passengers who are crowded and packed together between decks like sheep for the slaughter, in a confined atmosphere, daily becoming more noisome, without the means of ventilation or disinfection. Can any sane man say the disease—the cholera—was not here, on board these ships, generated and produced?

This is also confirmed by the occurrence of an isolated case on Ninety-third Street, near Third Avenue, the first case in this city this season. Though the cholera exists at Quarantine, the patient had not been in any way exposed to the disease, except to the exhalations from the overflowing and drainage of a privy and the foul atmosphere arising from the cellar of her own tenement. On Monday, it is said, she partook of her dinner, feeling a little indisposed; at 4 P. M. she called in her physician, and died the next morning, May 1st, at 11 A. M., in a state of collapse.

Take another instance: the second case in this city occurred in one of the tenement dwellings of the Sixth Ward, No. 117 Mulberry Street. The patient was a woman about thirty years of age, who had not been exposed, except to the noisome atmosphere of her own dwelling and its surroundings, which must be regarded, under the peculiar circumstances, as a true, genuine cholera atmosphere. In these cases the evidence is conclusive that the disease was generated and produced within, and on these premises.

The exhalations from low, moist, and marshy localities, from the offensive cesspools, water-closets, sinks, sewers, and the decomposition of animal and vegetable substances, from the refuse or garbage which so often befouls the sidewalks and gutters of streets, are all effective, predisposing causes, that directly facilitate the production of the cholera. Whatever tends to depress the vital powers, impair normal action, or relax in any degree the tone of the nervous system, favors the operation of the final cause. So, too, the low, underground, damp, unventilated apartments, the crowded and uncleanly tenement houses, in which multitudes of the poorer class live, in a confined, foul, and noisome atmosphere, not only favor, but actually invite, the active operation of the infecting agent.

Habits of intemperance, profligacy, impurity, and late hours, have a powerful influence to depress and prepare the system for an invasion of the disease in its most malignant form. In a neighborhood of this description, when the cholera in 1832 was raging in the adjacent city, from which it was separated by a very small creek, the uncleanly multitude escaped entirely, not a case occurring there at that time; but when, after an interval of several weeks, all danger seemed to have passed, and the people were rejoicing and congratulating themselves on their good fortune, the fearful disease suddenly appeared in their midst with greatly intensified effect, and in a very few days swept the place so clean that few were left to tell the sad story of its ravages.

There are some other predisposing causes of no inconsiderable influence, which not only favor the operation of the infecting agent in the production of the disease, but even awaken its latent power, and stimulate its activity and development in the system, once exposed to its invasion. Among these, excessive fear of an attack, great anxiety and depression of mind, constitutional debility, deranged condition of the digestive organs, accompanied with a relaxed state of the bowels, exhaustion arising from fatigue or disease, semi-starvation and unwholesome diet, neglect of personal and domestic cleanliness, irregular habits, and excesses of every description, are all direct incentives and stimulating agents in the production of the cholera. Any one of these may be sufficient to induce an attack; but when a number unite and act conjointly the danger is vastly greater, as the infecting agent or disease-poison becomes thereby more intensified.

When the cholera first appeared in Europe and in this country in its epidemic form, the majority of medical men, as well as the people, believed it to be contagious, and to be propagated solely on this principle. But when the disease appeared in 1848 a decided change of opinion occurred, which led to a full discussion of the subject, without any definite result; and the great question as to its contagious character and its mode of propagation remains still unsettled. The higher authorities, says an eminent author, concurred in the opinion of the Board of Health, " that the disease was not in any way contágious, and that no danger was incurred by attendance on the sick."

A large body of evidence, however, has been exhibited, going to show that human intercourse has, at least, a share in the propagation of the disease, and that it, under some circumstances, is the most important, if not the sole means of effecting its diffusion. On the other hand, it is affirmed that though it may be communicated, in some cases, by the agency of human intercourse, it does not follow that the material cause spreads by true contagion, that is, by reproducing itself in the bodies of men, and there only.

The disease may be carried by healthy persons in their clothing, in their ships, and in their caravans. That instances of this kind have occurred there can be no question, for numerous records present some undoubted instances of the occasional communication of the cholera-poison through human intercourse; still it is no less certain that its general extension over the world cannot be accounted for on this principle alone. "Its propagation by this means seems to be the rare exception, its spread over the earth from other causes being the common rule."

Dr. Hamlin, writing from Constantinople, in reference to the recent irruption and prevalence of the cholera in that city, observes, "The idea of contagion should be abandoned. All the missionaries who have been most with the most malignant cases, day after day, are fully convinced of the non-contagiousness of the cholera. The incipient attacks which all have suffered from are to be attributed to great fatigue, making the constitution liable to an attack."

It is a very singular fact, that the medical profession in India, the birthplace and home of the cholera, almost universally reject the doctrine of contagion. If those most observant and familiar with its history, its prevalence, and its annual recurrence as an endemic disease, which they are called to treat in all its varied phases, have discovered no contagious character by which it can be propagated, it may be safely inferred that it is not contagious in the common acceptation of the term, and that its extension over the earth is governed by some other principle, and that the predisposing and localized causes which are always in operation in India exercise no small share in its diffusion, in directing its course, aggravating its severity, increasing or diminishing its fatality, and determining the duration of its prevalence in particular localities. When its infecting germs have gained a lodgment in any city, section, or country, they may be stimulated and become exceedingly active in the production of the disease through these influences.

As to its introduction into different countries, it is quite evident that the germ, or latent principle of the cholera-poison, exists in such a state as to be capable of transportation, and may in this way be diffused to almost any extent when the localizing influences are sufficient to develop its energies.

In this, as in all other zymotic diseases, some persons are more susceptible of an impression and more liable to an attack than others. Though no class can be considered exempt, yet there are some whose organization, or innate protective principle, seems to render them impervious to its influence. The cholera, however, is no respecter of persons, or rank, or condition. The anæmic and cowardly in all ranks and conditions are peculiarly liable, and are the most defenceless and unresisting when invaded. In Europe, the probable numbers attacked in that part of the world appear from statistics to be, in France, as 1 in 300; Russia, as 1 in 20; Austria, as 1 in 30; Poland, as 1 in 32; Holland, as 1 in 144; Germany, as 1 in 700. "The circumstance of one attack by no means protected the individual from a second in the same, or any subsequent year; still a repetition of the disease in the same person in the same year was rare."

CHAPTER II.

Section I.—Pathology.

The doctrine now universally accepted and prevailing regarding its Pathology is, that a poison, virulent, subtle, and unknown, has been absorbed, and primarily infects the blood, so that, after a longer or a shorter time, a primary disease of this vital fluid is produced, and that the poison undergoes an enormous process of multiplication in the living body of the cholera patient, as the direct result of this morbific process so established, and that changes are induced in the function of respiration directly consequent on this alteration of the blood.

This altered condition and rapid change in the life-sustaining principle of the blood, the loss of nerve-power, the impaired circulation and tendency to congestion, are the proper and distinguishing features of the disease ; and the term " Algide," first used by the French Pathologists, very happily describes one of the most remarkable and constant symptoms, namely, the diminution of animal heat. The loss of temperature and its consequent effects upon the circulation, depressing and prostrating the nervous power, impairing and paralyzing the respiratory organs, suspending the functions of the liver and kidneys, enfeebling the action of the heart, and causing the capillary vessels of the mucous tissues to expand and pour off the serous fluid from the blood and every muscle and tissue of the system, with great rapidity, essentially constitute the phenomena of the Cholera. The constantly increasing augmentation of the poison and its intensified effects measure the malignity, the violence, and the rapidity of the disease.

It is this multiplication, and the disturbance which attaches to it, that in each case constitutes the disease and destroys life. Of this fact the circumstantial evidence is abundant and conclusive, and may account in part for the violence of the disease in its first irruption in any particular locality.

The vomiting, purging, and cramps are now generally considered as secondary and non-essential phenomena, for numerous cases of cholera have occurred in every section where it has prevailed in its more violent and malignant form without exhibiting these symptoms. The poison was so potent, and its progress so rapid, that life was extinguished in a very short time. In its first irruption at Muscat, cases are reported in which only ten minutes elapsed from the first apparent seizure before life was extinct. Dr. Milroy, speaking of the violence and rapidity of the disease as it occurred in 1817, and again in 1845 and '6, at Kurrachee, observes, that "within little more than five minutes hale and hearty men were seized, cramped, collapsed, and dead." Instances of death taking place in two or three hours are extremely common. When it broke out at Teheran, in May, 1846, Dr. Milroy observes, that "those who were attacked dropped suddenly down in a state of lethargy, and at the end of two or three hours expired, without any convulsions or vomitings, but from a complete stagnation of the blood."[3] In many places during its prevalence in 1832, and subsequently in 1834, and in 1848 and '9, the rapid fatal character of the earlier cases was observed and reported as the most severe and hopeless. In various cities and villages in our own country, cases of this description were not unfrequent. In all these the destructive nature and rapid process of the disease was so depressing and overwhelming as to prevent any effort of the "vis naturæ" to resist its progress.

Hence, from the autopsy of those who have fallen victims to its baneful influence in the first stage, or within forty-eight hours of the attack, no alteration of structure in any organ or tissue has been discovered. But in those cases where death has occurred at a later period, some lesions and slight changes in the appearance of some tissues have been traced. The more important of these, illustrative of the characteristic effects of the disease, are, in brief, the following:

The follicular structure of the intestinal canal has been found slightly swollen, and the intestine partially filled with a turbid, inodorous, semi-diaphanous fluid, resembling thin starch, or rice-water, and is supposed to be the remains of that peculiar secretion which had taken place during life. This fluid is sometimes acid, and sometimes alkaline. In the small intestines it is found in an unmixed condition. It consists of two liquids of different consistency; the one thick, the other thin. The latter constitutes the rice-water stools, and may be passed off without admixture with the thicker substance. The colon has been found generally much contracted, and the mucous membrane and the sub-mucous cellular tissue of the digestive canal presenting evident marks of congestion, in some cases approaching to a sub-inflammatory state, generally in spots or patches of various sizes, the color of these varying from a very dark congestion to a more roseate hue. The glands of Brunner and Peyer, as well as the solitary glands, are greatly enlarged.

The stomach and bowels are frequently of a paler color than natural, both in their inner and outer surfaces. The liver, the spleen, and the kidneys have been found engorged with blood. The urinary bladder is always contracted, and empty. The gall-ducts are sometimes contracted, at other times not. The vena porta and all the other abdominal veins are loaded with black blood, resembling tar in its color and consistency. The membranes of the brain and cord are generally found congested, and the substance of the brain more or less dotted with small points or specks of blood than usual.

"The most common appearances in the lungs," says an eminent pathologist, "are the presence of blood in the large vessels, chiefly or solely; the collapse and the deficient crepitation arising from the more or less complete absence of air and blood, and from the approximation of the molecular parts of the pulmonary substance. In other cases there is more blood in the minute structure, a corresponding dark color of the lung, and a variable amount of frothy serum. The right side of the heart and the pulmonary arteries were generally filled, and in some cases distended with blood; the left side and aorta were generally empty, or contained only a very small quantity of dark blood; the left side evidently had received little or no blood, but had continued to contract, in some cases even violently, on the last drop of blood which had entered it."

Such are some of the prominent appearances which the body has presented when the patient has died in the first, or pulseless stage of the disease. But in other cases, where the premonitory stage has been definitely marked, and attended with diarrhœa or other depressing disorder affecting the alimentary canal, and where the patient has continued under the influence of the disease for a longer period, and has passed through the usual successive stages of it, other additional appearances have been noticed, which are here omitted, as they are of a secondary importance, and belong especially to the more protracted cases.

5*

The *post-mortem* appearances, the phenomena of the disease, the Algide, or diminished animal heat, and the loss of nervous power, all tend to show an obstructed circulation and consequent embarrassment of respiration resulting in the non-aeration and non-oxydation of the blood, from which a long train of secondary and non-essential symptoms arise. For it is affirmed that the mechanical part of respiration remains in a good degree perfect, and that the heart evidently continues to beat in many cases till stopped by the want of blood in the left side and by its accumulation in the right side. Hence, for the cause of this arrest of the circulation of the blood through the lungs, we are forced to look to the condition of the blood itself, and the deranged action of the ganglionic nerves.

Attempts have been made to trace out from analysis the exact chemical changes in the order of their occurrence which attend the period of transudation from the blood into the intestinal canal. "The most prominent phenomena of cholera," says Dr.

Aiken, " during this period of transudation, consists in separation of the water and of the salts of the intercellular fluid (of the blood) through the mucous membrane of the intestinal canal, and the retention in the blood of an important excess of albumen and of blood-cells, with apparently less, but in reality with great diminution of the salts and fibrin."

"The inorganic constituents," continues the same author, "if compared to the water, are during the first four hours increased, because at this time the water is passing off with great rapidity; afterwards, as the salts pass off, the disproportion is lessened, and after eighteen hours or so, the proportion of salts is greatly diminished, and, if compared with the organic constituents, the diminution is enormous. With respect to the individual salts, there is in the blood a relative preponderance of phosphates over chlorides, and of potash salts over soda salts. By the end of eighteen hours or so, the blood-corpuscles are left in a most abnormal condition; the great loss of water and of salts, especially of the chloride of potassium—a most important constituent of the blood-cells—at once leads to the conclusion that their functions must have been greatly impaired. Accordingly, Dr. Schmidt found that the amount of oxygen contained in them was lessened by one-half." Dr. Robertson affirms that the "fibrin of the blood is usually in large amount and coagulable with great firmness;" while Dr. Parkes, speaking of the same condition of the blood, and relying on the accuracy of his analysis, observes, "The presence of fibrin in the blood was not indicated by any coagulation either in or out of the body; and whether coagulated or not, the blood has usually a dark color; but it generally acquired an arterial tint when brought into contact with the air in thin layers." * * * "When we remember the great share taken by the blood-globules in the respiratory and heat-furnishing processes, it is scarcely possible to avoid concluding that their loss of salts is connected with the characteristic cyanosis and lowered temperature in cholera." "The diarrhœa coincides with the first chemical changes in the blood—the transudation of some of the constituents of the serum." Hence the phenomena of the disease may thus be traced from this process as the starting-point. All other chemical changes in the blood, and the most marked symptoms, such as the abnormal respiratory process, follow as a matter of course. Such is the theory of the nature of cholera, now advanced and sustained by the most eminent pathologists, which embraces the doctrine previously advanced that the blood is the primary seat of the disease, and becomes contaminated by the absorption of a specific poison.

SECTION II.—PHENOMENA, OR SYMPTOMS.

The attack of this fearful disease is most generally sudden, the patient being at the time apparently unconscious of any depressing influence, or derangement of the system. It is not unfrequent, however, that some slight irregularity of the bowels, loss of animation and general vigor, or other apparently trifling indisposition, have preceded it. In some instances there are definite and decided premonitory symptoms which continue for a longer or shorter time prior to the attack, commencing usually with a pallor or collapse of the countenance, depression of spirits, slight pain in the forehead, noise in the ears, occasional or transient turns of vertigo, slight nausea, heat and pain in the epigastrium, oppression at the chest, with frequent sighing, nervous agitation, some loss of muscular power, general uneasiness, flatulence, with slight diarrhœa, sickness at the stomach, occasional twinges of the nerves, or cramps in the extremities, oppressed, small, feeble, and sometimes intermitting pulse, coldness, clamminess, or humidity of the surface, and general lethargy. Such are some of the premonitory symptoms which more frequently occur in the lower latitudes, where the general vigor becomes depressed by the long-continued and excessive heat of the climate. Their duration, whenever any of them do occur, varies materially; sometimes one, two, or three days—sometimes longer but not often.

According to the observations and descriptions given by those who have had the best opportunities for becoming familiar with all its various phases, the symptoms attending its invasion and general course are too distinctly marked to be ever mistaken for any other disease. In the minds of many who have been called to witness the developments of cholera, they undoubtedly exist with such distinctness and vividness as to render the most labored and accurate description tame. In this treatise, however, a description of the leading and more prominent phenomena will be given, and so far as a general principle of practice is concerned, this might be very appropriately limited to its first or cold stage.

The commencement of the disease is often so insidious as to pass unnoticed till the system is fully prepared for the sudden and violent attack. The slight, painless diarrhœa, depression of the nervous power, and occasional vertigo may all pass unheeded, and the patient be apparently in perfect health. He may retire to rest entirely unconscious of approaching danger, and after enjoying a sound and undisturbed sleep for hours, be, on awakening from his slumbers, seized with a remarkable sickness, perhaps vomiting, accompanied with most remarkable and profuse discharges from the bowels. These inordinate evacuations are usually attended with severe pains, extending down the thighs, and a sense of complete and almost perfect exhaustion. The physical powers and vital energies are immediately prostrated. The temperature rapidly sinks below the normal standard—the body becomes benumbed with an icy coldness—the skin becomes shriveled up, and almost insensible to hot and stimulating fomentations—the breath, too, as it comes from the lungs, appears to partake of the same icy coldness, indicating the rapid elimination of heat, or caloric, from the body. The patient complains of being greatly oppressed, throws off his clothing—calls for cold water, which he eagerly and copiously drinks; though it afford no relief to his insatiate thirst, it ought not to be withheld. This peculiar icy coldness and loss of temperature is also further shown by the livid, blue, or purple appearance of the hands and feet, extending sometimes over the greater part of the body. The skin becomes, even in a few minutes after the seizure, not only shriveled up, but often curiously wrinkled, as in extreme old age. Severe spasms in the fingers, toes, legs, and abdomen, cause the patient to groan and writhe under their influence, and to call on his attendants, if fortunate enough to have any around him, for aid and relief from his agonies. As the disease proceeds, there may be noticed a peculiar, sharp and contracted state of the features, and a wild and terrified expression of the countenance, arising from the impression and fearful apprehension of rapidly approaching dissolution. These important changes may all take place in a very few minutes. To these most obvious and singular symptoms there is superadded constant vomiting—incessant purging—low, feeble pulse, though occasionally natural and sometimes rapid, yet in some instances, from the very first moment of attack, cannot be discovered either in the large superficial arteries or at the wrist. The voice is altered, becomes low, feeble, unnatural in tone, or sinks even to a whisper. Respiration becomes quick, irregular, laborious and imperfect. The inspiratory act being performed with difficulty, and expiration being quick and convulsive. The flow of bile into the intestines is suspended, the urinary secretion and micturation entirely suppressed. Almost the only organ which seems to preserve in any good degree its powers is the brain—the mental faculties in some cases being retained till the close of life;—in other cases feeble, weak, and much impaired. On the accession of the spasms, the vomiting—and the purging—the disease may be considered as being fully developed, and the crisis at hand, which, in a few hours, must decide the fate of the patient. Its progress is now rapid, and must speedily terminate either favorably or unfavorably. If the result be unfavorable, the patient may die with all these symptoms distinctly and strongly marked. If the termination, however, be favorable, these violent symptoms soon yield, and seem to be materially relieved; yet, though these indications favor the return of normal power—the weakness, the cessation of the pulse, the coldness and blueness of the surface, and the sepulchral expression of the countenance, clearly show that a few hours must close the scene. To many death thus often comes calmly and quietly, without any struggle to mark the precise time of this life's departure.

"If the patient," says an eminent author, "should happily survive the cold stage, the disease may terminate by a rapid recovery, or it may pass into the second or febrile stage. The former is the more usual course in India, the latter in Europe. The first symptom of returning health is shown by the patient falling into a sleep of unusual soundness, during which the respiration becomes light and easy, the pulse freer, while a gentle, warm perspiration bedews the whole body. This grateful pause in the disease appears to be the result of the returning powers of life, uninfluenced by medicine, for it often occurs where none has been given. After this balmy slumber the patient awakes refreshed, and often recovers so rapidly, that in the natives of India it almost resembles a restoration after syncope. In all the Presidencies, indeed, and especially in Bengal, the recovery of the European has, in general, been followed by a stage of reaction, usually slight, but in some cases assuming the form of the bilious remittent fever of the country, which has occasionally terminated fatally. In most cases, however, the reaction is more considerable, and the patient, in a few hours after the subsidence of the cold stage, labors under a severe form of fever, resembling the typhoid. During the first few hours after the febrile reaction commences the tongue is white, but it quickly becomes brown and dry, while black sordes incrust the teeth and lips. The eye becomes deeply injected and red, the cheek pale or flushed, the pulse rapid, and the temperature of the body a little above the natural standard. The patient, either delirious or comatose, then lies in a state resembling the last stage of the severest typhoid fever of this country. This struggle usually lasts from four to eight days, when the symptoms either gradually yield, or death ensues. In a few mild cases the fever assumes an intermittent type, or sometimes a quotidian, sometimes a tertian form : all these cases usually recover. Such is, in brief, a summary of the more important symptoms of the Epidemic, or Asiatic Cholera, especially in its earlier or cold stage. The phenomena, especially developed in, and belonging to, the stage of reaction, being of minor importance, they have received only a very brief consideration; sufficient, however, to show the general character and tendency of the disease in this stage of its progress and termination.

CHAPTER III.

Section I.—Unsuccessful Modes of Treatment—Venous Transfusion Explained.

In this discussion we shall avail ourselves of the researches and investigations of eminent Professors, whose observations, experience, and position give their views the highest authority. The latest and most deserving record on this subject is from the pen of Professor Aiken, of Edinburgh, who observes, "There are few diseases for the cure of which so many different remedies and modes of treatment have been employed as in Cholera, and, unfortunately, without our discovering an antidote to the poison. In Moscow it is said that the mortality was not greater among the destitute of medical aid than among those who had every care and attention shown them. It may be fairly

UNSUCCESSFUL MODES OF TREATMENT. 65

inferred, therefore, that in the severer forms of the disease, the action of this poison is so potent as to render the constitution insensible to the influence of our most powerful remedial agents. When, however, the disease is mild, or on the decline, much may be done by obviating symptoms to promote the recovery of the patient."

" The heroic remedies that have been employed in Cholera are bleeding, and calomel and opium, either separately or conjointly. With respect to bleeding, it may be stated, that in every country the patients bore bleeding badly in any stage, and that the practice in Europe was at length limited to a few leeches occasionally to the head. As to calomel, that medicine was used to the greater part of an ounce in the twenty-four hours, but with so little success that many patients have been seized and have died under the full influence of mercury. On the appearance of cholera in Europe, opium was administered in the doses recommended by the Indian practitioners to the greater part even of an ounce of laudanum ; but it was soon seen that in the cold stage it was inefficient in controlling the vomiting or purging ; that it did not allay the spasms, and, moreover, hardly produced any narcotic effect. The action of the accumulated doses of opium, however, though suspended during the cold stage, was often fully developed in the last stage, and occasioned so much affection of the head that most practitioners either abandoned its use or limited it to a mere fractional dose of that usually given in India, namely, from three to twelve minims of the tincture of opium, or half a grain to a grain of solid opium every four or six hours."

Let us now turn to a paper by the justly celebrated Professor Maclean, whose observations and experiences have been more extensive than perhaps those of any other professional gentleman either in Europe or America. Unlike many of his brethren, he holds on this subject the safer doctrines of practice, and very frankly and earnestly expresses the same in the following language : " Opium in cholera should be given only in the premonitory diarrhœa. At this stage, in combination with a stimulant, it is of the highest value. If persevered in, particularly in the strong doses (justly reprobated), it is a dangerous remedy, inducing fatal narcotism, or, at the least, interfering with the functions of the kidneys, and so leading directly to uræmic poisoning."

" Urgent thirst is one of the most distressing symptoms in cholera; there is incessant craving for cold water, doubtless instinctive, to correct the inspissated condition of the blood, due to the rapid escape of the liquor sanguinis. It was formerly the practice to withhold water—a practice as cruel as it is mischievous. Water in abundance, pure and cold, should be given to the patient, and he should be encouraged to drink it, even should a large portion of it be rejected by the stomach; and when the purging has ceased, some may, with much advantage, be thrown into the bowel from time to time.

In the stage of reaction, the fever may be moderated by cold sponging, or by the wet sheet ; the secretion of urine may be promoted by dry cupping over the loins by the use of chlorate of potash, and the like. But suppression of this secretion is most to be dreaded where opium has been too freely used in the treatment. In men of intemperate habits, we often see, during the stage of reaction, obstinate vomiting of thick, tenacious, green, paint-looking matter, probably bile pigment, acted on by some acid in the stomach or alimentary canal. It is a symptom of evil omen, and often goes on uncontrolled until the patient dies exhausted, and this although all other symptoms may promise a favorable issue. I have known it last for a week, resisting all remedies, and proving fatal when the urinary secretion had been restored and all cerebral symptoms had subsided. Alkalies in the effervescing form, free stimulation of the surface, and chloroform in small doses offer the best hope of relief. The patient should be nourished more by the bowel than the stomach when vomiting is present. Ice should be given *ad libitum* when it can be obtained, not only to dissolve in the mouth, but to swallow in pieces of convenient size."

"Another heroic plan," says Dr. Aiken, "peculiar, perhaps, to this country, which was practiced when the inefficiency of medicines was generally admitted, was an injection of a solution of half an ounce of muriate of soda, and four scruples of sesquicarbonate of soda in ten pints of water, of a temperature varying from 105 to 120 Fah., into the veins of the suffering patient. The solution was injected slowly; half an hour being spent in the gradual introducion of the ten pints, and the immediate effects of this treatment were very striking. The good effects were rapid in proportion to the heat of the solution, but a higher temperature than what is stated could not be borne. After the introduction of a few ounces, the pulse, which had ceased to be felt at the wrist, became perceptible, and the heat of the body returned. By the time three or four pints had been injected the pulse was good, the cramps had ceased, the body, that could not be heated, had become warm, and instead of cold exudation on the surface, there was a general moisture; the voice, before hoarse and almost extinct, was now natural, the hollowness of the eye, the shrunken state of the features, the leaden hue of the face and body had disappeared, the expression had become animated, the mind cheerful, the restlessness and uneasy feelings had vanished, the vertigo and noises of the ear, the sense of oppression at the precordia had given way to comfortable feelings; the thirst, however urgent before the operation, was assuaged, and the secretion of urine restored, though by no means constantly so. But these promising appearances were not lasting; the vomiting continued, the evacuations became more profuse, and the patient soon relapsed into his former state, from which he might again be aroused by a repetition of the injections; but the amendment was transient, and the fatal period not long deferred. Of 156 patients thus treated at Drummond Street Hospital, Edinburgh, under the direction of Dr. Macintosh, only

25 recovered; a lamentably small proportion; and, small as it is, it seems doubtful if the recoveries were final or complete."

But let us turn to another page, whose beauty is especially marred by unreasonable expedients: "The warm bath," says the writer, "was at first tried, but discontinued from the uncontrollable nature of the vomiting and purging, and the oppressive sensation of heat it produced on the patient's feelings. Mr. Dalton's vapor bath and Turkish baths in the Hospital at Scutari have been used, but without benefit, and to the disappointment of the hopes which had been entertained of them."

Other methods of restoring warmth were had recourse to, such as frictions with the hand or by the flesh-brush, or rubbing the body with some strong stimulant embrocation, compounded of garlic, capsicum, camphor, cantharides, or other powerful irritants. Mustard poultices also were often applied to the feet and abdomen, blisters with or without an addition of oil of turpentine, the part having been previously rubbed with hot sand; and in cases supposed to be urgent, the mineral acids, and even boiling water, were employed for the purpose of producing instant vesication."

"And, again, we read of those who tried to stimulate the waning powers of life by galvanism, acupuncture of the heart, issues, setons, moxas, actual cautery along the spine, and, lastly, by small pieces of linen dipped in alcohol distributed over the body and then set fire to!!!" Such are some of the means which have been used in the treatment and cure of cholera.

"The failure of such powerful means at length caused most practitioners to confine themselves to checking the diarrhœa, which so frequently precedes cholera, and subsequently, to obviating symptoms as they arose," and for this purpose, returned to and adopted a very simple stimulating mixture, recommended by the Board of Health:

℞. Pulveris Aromat., . . . ʒiij.
    Tinc. Catechu, ... " x.

Tinc. Cardamom, C., . . ⸪ vj.
Tinc. Opii, . . . ⸪ ⸪ ⸪ j.
Mixt. Cretæ Preparat., . . ℥ xx.
M.—— S., j ℥, as necessary.

Tinc. Kino, or the decoctum Hæmatoxyli, were sometimes added.

These remedies, it is said, frequently arrested the attack altogether. If, however, the disease proceeded and the cold stage of cholera formed, the same remedies were prescribed in an effervescing draught. "To promote reaction in cholera and diarrhœa, the following formula has met with most universal approval in this country and in India. So highly is it valued, indeed, that it is ordered to be always in store, and in readiness in the *Medical Field Companion* of the army when on the march :"

℞. Ol. Anisi, ⎫
   Ol. Cajeput, ⎬ āā. ʒ ss.
   Ol. Juniper, ⎭
   Æther, ... ℥ ss.
   Liquor Acid. Haleri,* . . ʒ ss.
   Tinc. Cinnam., ... ℥ ij.

* Sulphuric acid, one part; Rectified Spirit, three parts.

M.——— S.: ten drops every fifteen minutes, in a table-spoonful of water. An opiate may be given with the first and second dose, but should not be continued."

The learned author to whom we have referred, after detailing some of the various expedients employed in the treatment and cure of cholera, sums up the whole under the common term—failure—and, in effect, declares the most powerful remedial agents ineffective and useless in controlling and subduing this disease.

This declaration is made in reference to the general result of the remedies and the various expedients adopted mainly by one class of physicians, to which special reference has been made. It is therefore partial, and confined solely to what is erroneously termed the regular practice. In declaring all remedial agents a failure, does not the author himself commit a greater failure in omitting to survey the whole subject of treatment, and to trace out and to show from the application of the pathology of the disease the probable cause of such failure?

However formidable this disease may appear, on account of its rapidity and its firm, unyielding grasp upon the vital powers, the forbidding and almost hopeless prospect of relief, and the lamentable results which have attended some modes of treatment, it seems particularly unfortunate for the profession that there should have been a disposition on the part of this learned author to abandon all remedial agents as comparatively useless, without a more thorough investigation into the cause of failure. On this point no effort or inquiry even is made. This is the more remarkable and surprising after dwelling at length on the pathology of the disease. It would seem as if all the light and science derived from this source for nearly half a century had been overlooked, or the pathology of the disease, from some cause not satisfactorily explained, had been deemed unworthy at least in this instance to dictate the course of treatment. This should govern in cases of cholera as in all other forms of disease, or else all our efforts and remedies will prove abortive. Now, had the doctor carefully investigated the various modes of treatment and compared the results of each, he might have come to a different conclusion. But, being confined and limited in his investigations, he is unable to discover anything reliable or worthy his commendation, except the formulas above and the recommendation of Dr. Maclean. Among all the remedies and expedients named, there is only one tending to fulfil, the indications required, and that one, though prompt and magical in its effects, has been unequivocally condemned, without looking beyond the transient result for any light it might shed upon the subject. How it should have escaped his notice and passed so long unobserved by the numerous professional gentlemen who had often witnessed the effect, and were anxiously searching for light and the means of affording relief to the suffering patient, is a most singular circumstance which can only be accounted for on the principle that they all were anticipating some strange phenomenon, or development of cure as mysterious as the disease itself, which led them to overlook the simple and effective means of relief so clearly represented and shown in their numerous experiments for something more heroic and powerful than as yet the imagination ever conceived.

If we trace the action of calomel, the use of opium, the effect of cupping, bleeding, blistering, etc., etc., we shall obtain no very desirable information; nothing valuable tending to indicate a correct principle of practice. If we go still further, and examine the tendency and effects of the various baths exhibited at Scutari, the use of the flesh-brush, the bare hand, the heated sand, the embrocations, the turpentine and other irritants, the boiling water, or the burning alcohol, skinning and cooking the patient alive, we shall be shocked at the enormous cruelty and barbarity that have been pursued, and turn from the repulsive exhibition, without discovering one ray of light to guide us in the right direction. Disappointed and baffled in our inquiries, shall we here abandon our investigations and dismiss the whole subject, because our course is involved in difficulties? Would intelligence and reason justify the neglect to improve the means at command? We think not; but rather induce us to advance in search of truth if the elements of success are not quite exhausted. Let us be encouraged and stimulated to untiring perseverance so long as there remains any experiment untraced and uninvestigated in its bearing upon the direct action of the disease. Had Dr. Aiken, or those other eminent surgeons who took part in those numberless experiments, instituted on the Continent and in England, especially those who initiated the process of injecting into the veins a solution of soda raised to a temperature from 105° to 120° Fahr., continued their investigations patiently and assiduously, they might probably have discovered long ago the correct theory of practice for the treatment and cure of cholera.

But they failed to see, or, if they saw at

all, rejected the feeble ray of light struck out by the experiments in which they had themselves participated, and like the celebrated Dr. Hunter, who refused to listen to the discoveries made by his pupil, the indefatigable Jenner, who traced the identity of the variola with the common disease affecting the kine; and thence extracted the vaccine lymph and established a principle by which that loathsome disease and often recurring epidemic has been nearly banished from the earth. Though they have thus failed, they have nevertheless left on record, in unmistakable language, the result of their bold experiments, which we may investigate, and appropriate the instruction drawn thence for our own and the advantage of our fellow-men.

What, then, are these results, regarded as shedding light on this intricate subject? We refer only to one the most obvious which we have already cited above. Let us repeat and analyze, and, if practicable, show the principle evolved. There was, on various occasions, the solution of soda injected into the veins at the temperature from 105° to 120° Fahr. : a higher temperature could not be borne. This process was performed slowly, thirty minutes being occupied in injecting the ten pints. Now mark the result as the operation proceeds. Says Dr. Aiken, " After the introduction of a few ounces, the pulse, which had ceased to be felt at the wrist, became perceptible, and the heat of the body returned." Mark the language : " only a few ounces" were required to arrest for the time being, the progress of the disease and restore warmth to the body; a very remarkable fact, replete with instruction, as will appear as we proceed. Again says the Dr., " by the time three or four pints had been injected the pulse was *good*, the cramps had ceased, the body, that could not be heated, had become warm, and instead of cold exudation on the surface, there was a general moisture. The voice, before hoarse and almost extinct, was now natural; the hollowness of the eye, the shrunken state of the features, the leaden hue of the face and body had disappeared ; the expression had become animated, the mind cheerful, the restlessness and uneasy feelings had vanished ; the vertigo and noises of the ear, the sense of oppression at the precordia, had given way to comfortable feelings ; the thirst, however urgent before the operation, was assuaged, and the secretion of urine restored, though by no means constantly so." Such is the astonishing result obtained by this experiment, and this, too, when only three or four ounces had been injected—all the urgent symptoms mitigated and relieved. What, we ask, could have been more satisfactory, or better calculated to aid the discovery of an important truth? Every distinctive and fatal symptom for the time is relieved, and the normal condition and functions of the system restored ; a result which could only have been obtained by the evolution of a principle of sufficient promptness and power and diffusibility to arrest and utterly suspend for a time the force of this disease.

What, then, was the principle evolved in

this experiment, which gave immediate relief? Did it consist in the half ounce of muriate of soda alone, or in the four scruples of sesquicarbonate of soda, alone, or in the ten pints of water alone, or in the whole combined, or more especially in the high temperature to which the solution was raised? It is a well-established fact that, in order to raise the temperature of cold water to blood heat and above, a large amount of free caloric must necessarily be absorbed, and exist mechanically in the fluid; and, in this condition, the solution was introduced into the veins, and there evolved its vast amount of free caloric, which immediately permeated every organ of the system, arresting disease, raising the temperature of the body, and restoring its normal functions. Of this there can be little doubt. For free caloric is one of the most prompt, effective and diffusive stimulants known, and was evidently in this case the remedial agent which produced the result. True, it may be said the effect was transitory, and passed off as soon as the caloric became eliminated. This, however, cannot alter the nature, character, or influence of the principle on which it was produced. It is usually admitted that a remedy that has power to control disease, will, by its continued action and influence, restore the normal condition of the system permanently, or at least aid Nature to repair her own work. By this we would not be understood as advising a repetition of the experiment under consideration, even under the most urgent circumstances; far otherwise would be our advice. We are arguing for the purpose of evolving and establishing a general principle of practice.

The great question, then, is, Did the principle evolved fulfill the indications required? and if so, is it available and consistent with the pathology and the peculiar phenomena, or symptoms of the disease? To settle this point, we need only turn to the law and the testimony, the very highest authority on the subject. The doctrine now universally accepted and prevailing regarding its pathology is, that a poison, virulent, and subtle, and unknown, has been absorbed and infects the blood, so that, after a longer or shorter time, a primary disease of this vital fluid is produced, by which the vital energy is impaired, and all other morbific changes induced. The term "Algide," first used by the French Pathologists, very accurately describes one of the most remarkable and constant symptoms, viz., the diminution of animal heat. On this depend the altered condition of the blood, the depression of the nervous power, the impaired functions of the respiratory and all the vital organs which are essentially involved by the disease. The icy coldness of the surface, the breath, the extremities and general loss of temperature, all show the character of the disease and the wants of the system.

Did, then, the principle evolved accord with the pathology and phenomena of disease? And did it fulfill the indications required? If not, we ask by what means was the disease arrested, and all the urgent symptoms mitigated and relieved, or by what were the good effects produced, and the normal action for a time restored? Can the result be reasonably accounted for on any other principle than the one assigned—the stimulating power of the free caloric? We think not; for it accords most perfectly with the pathology and the peculiar phenomena of the disease. It assuaged the more urgent symptoms, answered the imperious demand of the waning powers, revivified and reinvigorated the vital energies, and restored for the time the normal tone of the system. What more could be desired in any single agent than the result here obtained? That it accomplished all this, there can be no question, according to the statement of the learned professors who have repeatedly witnessed and described the results.

The question, however, will arise, Can this principle be rendered available? Most certainly it can; and though it may not be convenient to introduce free caloric into the stomach, we can, by combination, introduce a stimulant of equal potency which shall be equally as prompt, effective and diffusive in its action, similar in its influence, and similar in its results. It is the principle—not the precise element for which we contend.

It is universally admitted that in many instances we may learn much from observing the manner of death which, in a majority of these cases of cholera, may be described by the term asthenia—a death similar to that which occurs in congestive fevers, and in some cases of accidental poisoning. Perhaps the most striking fact observed in these cases is the perfect exhaustion attending the last moments of existence, and the quiet, undisturbed manner in which life terminates. This very clearly shows the exhausting nature and congestive character of the disease, and gives us an idea of the course of treatment necessary to be pursued. If, then, there is anything to be learned from this source relative to its treatment, it does most certainly corroborate and strengthen the position we have here taken.

Another feature of the case in aid of our

UNSUCCESSFUL MODES OF TREATMENT. 87

position consists in its entire accordance with the modes of treatment which have been most successful in the cure of cholera. The two formulas cited above, and now most universally adopted in Europe and India, are based on a similar principle. So in this country 1832,-33 and 34, the successful modes of treatment consisted in the adoption of a principle essentially similar. Hence we infer from the teachings of this experiment, and from all the collateral facts on the subject, that the general principle to be observed in the treatment of this disease is a prompt and diffusive stimulant; and hence we deduce the doctrine already apparent, that every form of treatment, to be successful, must be based on a prompt and effective stimulant of sufficient power to meet as speedily the indications required, as did the free caloric in the experiment to which we have referred.

Here we might pause for a moment and examine the suggestion and doctrines advanced by the learned Drs. Bell, Johnson, and many other eminent practitioners in

India and Europe. We might further investigate the principles and trace the practical philosophy of such eminent surgeon, as Drs. Mackintosh, Thompson, Wallis Maxwell, Massy, Hill and Brady, all of whom have had opportunity of investigating the nature and character of the disease and extensive experience in its treatment.

We might also, in a further examination of the subject, embrace a host of American authors whose works teem with every shade of doctrine, and almost every variety and description of practice, some evincing a degree of skepticism on the subject more wonderful and marvelous than is becoming the great apostles of medicine. It would seem as if the guiding light of science and experience had forsaken them in this, the hour of their need; that facts and arguments had failed to illumine their minds, or direct their inquiries in the proper course for the discovery of "the truth." Their conclusions on this subject are, therefore, marvelously inconsistent and conflicting. Over this mass of specious and conflicting testimony we might long ponder, without deriving any very valuable information worthy an elaborate effort, or making any discovery to aid in the establishment of a general principle of practice for the cure of cholera. But this investigation must be deferred to another occasion, when time may permit a more thorough and critical examination of their doctrines and practice than can be presented in this brief essay. We would, however, remark in passing, that in some instances their philosophy, doctrines, and results may lead us to the same conclusion to which we have arrived from other sources as above, and from our regard and belief in the progress of science, feel compelled to advocate the same, as offering the best hope of success in the treatment of this disease.

In the employment of an anti-miasmatic principle and remedial agent, we feel ourselves abundantly sustained, by the concurrent testimony, of those English surgeons connected with the Medical Bureau in the departmnet of India, whose numerous experiments and carefully detailed clinic cases occurring in the recent irruption and prevalence of the disease in that section, exhibit its utility in such a striking contrast with all former practice, as to leave no doubt as to its direct and specific action in the cure of cholera. It is in allusion to these experiments, and in answer to the question, what is deemed the most successful mode of treatment, that the learned Professor Maclean unhesitatingly observes, "Alkalies in the effervescing form, free stimulation of the surface, and chloroform in small doses, offer the best hope of relief." As this opinion comes from such high authority, and is compatible with the pathology of the disease, we may, without fear of controversy, add in conclusion, in any and every form of medication for the cure of cholera, we must not forget that chloroform is our sheet-anchor; and must be so combined and administered as to meet promptly the indications required.

Section II.—Physiological Condition of the Blood.

Its Non-Aeration—Non-Oxydation.

In the preceding section we alluded to the suggestions and doctrines advanced by the learned Dr. C. W. Bell, Physician to the Manchester Infirmary, and late Physician to H. M. Embassy in Persia—and also to Dr. George Johnson, of Kings College, whose views and doctrines, relative to the Pathology, illustrative of the congestive character and non-aeration of the blood, coincide with those of Dr. Bell. A brief examination of their philosophy and doctrines will show very conclusively the first direct impression of the poison—the gradually altered condition of the blood, and the corresponding loss of nerve power—the impeded arterial circulation and the general tendency to congestion, as well as the altered condition and stagnation of the blood, especially during the stage of collapse.

The question is asked. "What is the pathological explanation of this remarkable train of symptoms?" and the answer is given, "The one great central fact is this, that during the stage of collapse, the passage of blood through the lungs, from the right to the left side of the heart, is in a greater or less degree impeded." Very conclusive evidence as to the existence of impeded pulmonary circulation during life is afforded by the appearances observed in the heart, blood-vessels, and lungs after death.

After adducing the evidence of this impediment from *post-mortem* examinations, and affirming that the blood does not flow freely through the lungs and pulmonary arteries, which are often filled and much distended with blood, it is observed—"The most interesting and conclusive evidence that arrest of blood in the lungs is the true key to the pathology of choleraic collapse, is to be found in the simple yet complete explanation which it affords of all the most striking chemical phenomena of the disease, the imperfect aeration of the blood, and the suppression of bile and urine."

And again, says the learned author, "It is obvious that the stream of blood from the pulmonary capillaries to the left side of the heart is the channel by which the supply of oxygen is introduced into the system. One necessary consequence, then, of a great diminution in the volume of blood transmitted to the left side of the heart must be, that the supply of oxygen is lessened in a corresponding degree. This position, probably, will not be disputed by any one who will give the subject a moment's consideration. Nor, again, can it be denied or doubted that certain results must of necessity follow this limited supply of oxygen." * * *

" The blood in cholera is black and thick only during the stage of collapse; in other words, during the stage of pulmonary obstruction and defective aeration."

Again, in his explanation of the injection of the solution of soda into the veins of the suffering patient, it is affirmed, " The benefit, however, is of but short duration, for the primary cause of the impeded circulation, namely, the poisoned condition of the blood, being still in operation, * * * the stream of blood through the lungs will soon again be obstructed, and the patient thus passes into a state of collapse as profound as, and more hopeless than, before. It appears, therefore, that the hot saline injection into the veins and the operation of venesection, when it rapidly relieves, as it often has done, the symptoms of collapse, have this effect in common, that they facilitate the passage of the blood through the lungs, and thus lessen that embarrassment of the pulmonary circulation which is the essential cause of choleraic collapse. But whereas the *hot injections act* by removing the impediment which results from spasmodic contraction of the arteries ; *venesection acts* by relieving over-distension of the right cavities of the heart, and thus increasing the contractile power of their walls."

Such are, in brief, the views of the learned Drs. Johnson and Bell, whose works are very highly commended by their American editor to the notice of the profession. These views, coming as they do from the highest authority, fully sustain the doctrine that the earliest impression of the disease is made upon the blood, and hence it becomes altered and changed in its most essential life-sustaining principle; for its oxygen becomes diminished, its consistency augmented, and its flow through the lungs impeded. Through this channel the effect of the poison soon makes an impression on the ganglionic mechanism, and the nerve-power becomes correspondingly diminished, and the action of the ganglionic nerves essentially deranged. But this is not all: they exhibit in the clearest manner the congestive character of the disease, and show the necessity of prompt and decided means to arrest this tendency. Hence, they urge, in the strongest terms, the importance of observing carefully this essential feature, and endeavor to exhibit fully the condition of this vital fluid at a particular stage of the disease, when bleeding, as recommended in their practice, is required, and may be performed to the best advantage for the relief of the partially congested blood-vessels, and to stimulate and give freedom to the circulation. The passage of the blood, they affirm, is impeded, clogged, and partially suspended. To remove this obstruction, relieve spasm, and secure the prompt aeration of the blood, in hope of arresting the progress of this disease, is ostensibly the object. However, they seem studiously to avoid the most logical conclusions of their explanations, and justify a practice that can give no hope of permanent relief, while every fact and symptom is ominously suggestive of the wants of the system, which imperiously demands the aid of electrified oxygen, ozone, or free caloric, for the oxydation of the blood.

Says Dr. Reid, " I believe the true explanation of the arrest of blood in the lungs to be this : The blood contains a poison, whose irritant action upon the muscular tissue is shown by the painful cramps which it occasions. The blood thus poisoned excites contraction of the muscular walls of the minute pulmonary arteries, the effect of which is to diminish, and, in fatal cases, entirely to arrest the flow of blood through the lungs."

Says Dr. Wallis, "The phenomena which are exhibited when the deleterious air has been drawn into the lungs are these: the great gastro-pulmonary nerve is either wholly or partially paralyzed; the consequences are the cessation of all its functions, either wholly or partially. This great nerve is a nerve of function, and performs the functions of digestion and respiration, and influences all secretions."

Dr. Maxwell, of Calcutta, uses the following language: "The development of the stages of fever entirely depends on the changes the *leaven has effected*. If this change has been such that the blood has become too thick to flow through the lungs, then, as a matter of course, the collapse stage is developed in excess; in other words, *cholera asphyxia* is exhibited. The blood, unable to pass through the middle passage into the arteries, collects and swells out the veins, giving that deadly or blue color to the skin. When the vomiting and spasms come on, this mass of blood in the veins is squeezed with great force, and hence the clammy moisture that is forced from every part during these fits. There is no pulse, because there is no blood in the arteries." "There are also lethargy and languor, and oppression in breathing, caused by the blood being collected in the veins. These make up the principal links in the chain of mechanical symptoms."

Dr. Bell, dwelling on this congestive character of the blood, and endeavoring to point out the best mode of relief, observes, " When this has reached to such a point as to oppress the action of the heart, yawning first and then shivering, or a sense of suffocation and pain in the precordia, are the indications of oppressed circulation, and of the commencing effort of the heart to overcome the mass of blood which is stifling it. If, by the application of tourniquets to the limbs, or by *bleeding*, part of the blood which is rushing from the extremities to increase

CONDITION OF THE BLOOD.

this congestion is prevented from reaching the great veins, the heart, excited to increased action, is enabled, by this relief, more quickly to overcome the obstruction and restore the balance of the circulation, and the paroxysm passes off. If not thus mechanically aided, the heart, after a severe struggle to maintain the circulation during the period of constriction, is at length relieved by this nervous disturbance or spasm of the capillary circulation passing off of itself, and then the heart and arteries, so long excited by the struggle, maintain for a time their increased action after the obstruction in the capillaries is removed, and produce apparent febrile action. Presently this excitement subsides, the vessels become relaxed, and sweat succeeds. The vessels continue in this state for a longer or shorter period, according to circumstances, till they at length recover their ordinary tone and action in the intermission. This fever, however, is not fever properly so called, but reaction; and the sweating not critical, or essential, but relaxation. The cold stage is alone essential, and is the physiological cause of the subsequent stages."

From the passages we have cited, it is quite evident that Drs. Johnson, Bell, Parkes, Reid, Wallis, Maxwell, Massy, and many others, admit this congestive character and impeded circulation of the blood to be the result, or consequent of a primary affection of the blood, as we have already observed in a former paper. The *term* "Algide" is peculiarly expressive of the diminished animal heat, and, as Dr. Bell represents, it is the cold stage which is alone essential, and is the physiological cause of the subsequent stages. It is the specific disease-poison, so often referred to, that has been inhaled, the leaven that has effected such obvious changes in the blood. The poison, virulent, and subtle, and unknown, so marvelously active in its operations, that is exhibited so prominently in all the works we have perused as the one great, mysterious, and efficient cause which produces the disease called cholera, and all the phenomena of its development. To its direct and spe-

CONDITION OF THE BLOOD. 101

cific action, therefore, must be attributed all the phenomena of the disease as the resulting subsequent consequences.

It is also further evident, from the pathological facts and arguments adduced in support of this theory of congestion, that the abnormal condition or state of the bloodvessels is the result and the product of the activity of the primary or final cause, and must be regarded in relation to it as cause and effect. On this principle alone, the thickening of the blood, the contraction of the left ventricle of the heart, and of the capillary and pulmonary arteries, assigned by some as the cause of choleraic collapse, must be accounted for. These effects are not and cannot be from a process independent and outside of the primary disease action, but are the result of such primary action.

Again, it is evident, from the views and doctrines cited above, that the disease is decidedly congestive in its tendency and character from its very commencement. The impeded flow of the blood—the comparative emptiness of the left ventricle of the heart and arteries—and the excessive loss of temperature, all indicate a rapid process of congestion attending the progress of disease. This is one of the peculiar and prominent features of cholera, and is strikingly exhibited in the morbid appearances observed in all those instances where death has occurred within a few minutes from the first indications of attack.

When the attack is violent, the process is rapid; when mild, it is slow; and even in the collapse stage progresses tardily. In either case it is the direct resulting consequent of the primary cause. How else can the violent attacks, suddenly terminating in death, be accounted for? To what other principle can this altered condition and stagnation of the blood be attributed? The evidence confirmatory of this position is abundant and conclusive. Many instances of the apparently rapid action of the cholera poison are related by Dr. Milroy, in a historical sketch of the epidemic of 1817; and at Kurrachee in 1855 and 6, it is said, that within little more than five minutes, hale and hearty men are seized, cramped, collapsed, and dead!!

When the disease broke out at Teheran, in May, 1846, Dr. Milroy states that those who were attacked dropped suddenly down in a state of lethargy, and at the end of two or three hours expired, without any convulsions or vomitings, but from a complete stagnation of the blood.

In the paper before us, it is stated, that "in a great majority of cases in which death has occurred during the stage of collapse, the right side of the heart and the pulmonary arteries are filled, and sometimes distended with blood; the auricle being partially, and the ventricle completely and firmly contracted. The tissue of the lungs is, in most cases, of pale color, dense in texture, and contains less than the usual amount of blood and air. There is something surprising in the contrast between the almost constant occurrence of this extremely anæmic condition of the lung, from which scarcely even a few drops of blood flow when the tissue is cut, and the hyperæmia of most of the other viscera." This impeded flow of the blood through the lungs, resulting, as it must, in a very scanty supply of blood to the arteries, in connection with the corresponding fact of the increased expansion of the veins, filled with black, and thick, and stagnant blood which, by the action of a powerful poison, or malignant disease, has become disorganized and unfitted for circulation, furnishes indubitable evidence of one prominent and characteristic feature of cholera which we term congestion, and to which we alluded in our remarks when the question under consideration was first introduced; in this view we are happy to find ourselves, on a more thorough examination of the subject, ably sustained by eminent pathologists and authors, who have arisen during the half century last past, and whose works are said to embrace all that is known and reliable on the character and treatment of Epidemic Cholera.

It is worthy of notice, before passing

from this part of our subject, that according to Dr. Bell's *views*, the blood is forcibly sent into the great central veins, and there stopped in its course without any attempt to account satisfactorily for its singular arrest, at that point—Dr. Johnson comes to his relief, lifts the veil, and explains why it is kept there and cannot get any further. If the road, he tells us, had been clear and uninterrupted through the lungs, the blood would easily have got round to the left ventricle, and have again gone its round, but it is stopped by the spasmodic contraction of the minute branches of the pulmonary artery, which will not even allow the blood to enter the pulmonary capillaries, as shown by the remarkable anæmia of the texture of the lungs.

In this connection may be introduced an opinion as to the cause of the disease and some of its phenomena, which has obtained at least some celebrity, and attracted the attention, if not the careful consideration of the profession. It will account, in part, if founded in fact, for the physiological condition under consideration.

It is said, some have observed a chemical change in the constitution of the atmosphere, and have attributed the cause of the cholera to the loss or diminution of its ozone—a principle which is understood to represent what is very properly termed electrified oxygen. Ozone is, therefore, the vital element of the air. It is said that oxygen cannot be assimilated or combined with the blood except when it is in an electrified state constituting the peculiar property or state of ozone. In this state it produces vital electricity of the blood, *which is the life.* The brain is considered and represented as the reservoir of this vital electricity, and the nerves are the telegraphic wires or conductors of it. As a necessary consequence, all acts of material and intellectual life depend upon this double cause. The absence, then, it is affirmed, of this principle, termed ozone—or electrified oxygen—from the atmospheric air in certain localities and the consequent non-aeration or non-oxydation of the blood, may be considered as an efficient cause which will account for some of the most striking phenomena of the cholera.

Whether this electrified oxygen, or ozone, is identical with free caloric, it is unnecessary for our purpose at present to determine. It will be admitted that oxygen is the source of animal heat, and when introduced into the system generates its free caloric, which is an essential life-sustaining principle.

Dr. Massy, after describing a severe and advanced stage of cholera, observes, "The treatment of this case depends in the first instance on bleeding, and largely, if the patient's pulse is good, giving at the same time twenty grains of calomel with one of opium. This, he thinks, will be found the best practice. After twenty minutes, he gives ten grains more of calomel and half a grain of opium. He considers, however, a reliance on opium in this form of cholera most faulty—but observes, as you draw blood, stimulate, give punch, brandy, or wine and water, or carbonate of ammonia. Apply friction, with stimulating and hot liniments to the extremities, warm sand-bags to the feet, sinapisms to the calves of the legs and pit of the stomach ; for, if you can once raise the pulse, the chances in favor of recovery will be vastly increased." The practice of bleeding and stimulating at the same time is deemed of vast importance. Dr. Bell coincides in this view, and devotes much space to the necessary instruction as to the time when and under what circumstances to bleed and to what extent, endeavoring to show the advantages arising from a strict observance of certain rules in carrying out this practice.

We have thus traced, *in extenso*, the views and doctrines of eminent surgeons and authors on the changes of the blood, and especially of the impeded circulation, to show, if practicable, the inconsistency of the more common and prevailing practice, and its utter inadaptation to the pathology and phenomena of disease. On the latter there seems to be little or no discrepancy—on the former there is a great diversity—as there has been no general principle established and laid down as the basis

CONDITION OF THE BLOOD. 109

of treatment and cure of cholera. It has often been observed there is no disease on which so many different modes of practice have prevailed, some purely experimental, others empirical—and all without discovering an antidote to the poison, or any efficient mode of relief. The cause, or the poison producing the disease, still remains undiscovered. The direct mode of suspending and removing it, or counteracting its power and neutralizing its effect, and subsequently eliminating *it* from the system, remains still in doubt. What course, then, should the epidemic cholera again prevail in our midst, shall we pursue? Shall we rest satisfied with the diversified modes of treatment now prevailing? Or guided by the light of reason, science and experience, endeavor to adopt a general principle of practice, and exhibit and establish an efficient and judicious system, consistent with the pathology and the phenomena of the disease? Does then the practice, the prominent features of which are given above, accord with the indications required? In short, does the exhibition of bleeding and calomel and opium, accompanied with sinipisms, and hot, stimulating applications to the surface, meet the pathological condition and the phenomena of the disease? We have seen that the rapid changes in the blood, and the consequent direct tendency to congestion, are the proper and distinguishing features of the disease;—and hence the diminution of animal heat and general loss of temperature and their consequent effect, impeding the circulation, depressing and prostrating the nervous power —impairing and paralyzing the respiratory organs—suspending the functions of the liver and kidneys—enfeebling the action of the heart, and causing the capillary vessels of the mucous surfaces to pour off the serous fluid from the blood, and every muscle and tissue of the system with great rapidity, essentially constitute the phenomena of the cholera;—and that the constantly increasing augmentation of the poison and its intensified effects, measure the malignity, the violence, and the rapidity of the disease.

Is there, then, any tendency in bleeding to arrest this rapid process of disease so disorganizing, depreciating, and enfeebling to the vital life-sustaining fluid, the blood? Can abstracting a portion of it, however large, suspend the poison, or its activity, or even check its progress in its rapid course and fatal termination? Can it have, under its depressing and depleting process, any tendency or power to relieve the congestion that is taking place, or change in any good degree the poisonous principle which is now generally admitted to exist in the blood, and to be the sole and efficient cause of its altered character and condition? The poison, once introduced into the blood, like the leaven hid in three measures of meal, will continue its activity, increasing its energy, and multiplying its forces, till the whole circulation becomes affected, and its life-sustaining power is destroyed and utterly lost, unless, by the exhibition of some remedial agent, it shall be promptly arrested in its progress, and suspended and eliminated. Again we ask, Will calomel fulfill any of the indications required? Has it any influence or power to arrest this disease, to quiet the nervous system, relieve the cramps, or restore warmth to the body? Its specific action, so far as known, can have no tendency whatever to relieve the system in any essential particular, or stay the progress of disease, or delay its inevitable result, if it remainun subdued by the action of other remedies. Its action upon the liver, however prompt it may be, is only of a secondary importance. The primary cause must be overcome, its activity and energy suspended and the system generally relieved, or there is little hope in the case.

Here we may ask, Will opium aid, or give the relief so urgently demanded? However serviceable as an astringent and anodyne in the premonitory stage of the disease, it cannot be exhibited in the second stage to so good an advantage, as its direct influence is to aid and promote congestion in those cases, where a tendency of this kind is already in existence. Hence, its continuance in the true or collapse stage of cholera is now generally considered faulty.

Once more: The auxiliaries employed in aid of the leading remedies already noticed may be summed up in the language of the celebrated Dr. Massy, in his instructions and directions on the subject of the treatment now under consideration. He observes, "But, as you draw blood, stimulate, give punch, brandy, or wine and water, or carbonate of ammonia. Apply friction, with stimulating and hot liniments to the extremities; warm sand-bags to the feet, sinapisms to the calves of the legs and pit of the stomach; for, if you can once raise the pulse, the chances in favor of recovery will be vastly increased."

To these directions there can be no special objections, except in the first instance in which he, indirectly, commends the use of means tending to deplete and depress the system, already brought by disease to the very verge of utter exhaustion. Remedies of this tendency are contra-indicated, and cannot, to say the least, be employed to advantage.

Depressing remedies generally, instead of checking, or counteracting the disease, will inevitably aid and hasten its fatal termination. Stimulants, such as are prompt and diffusive in their character, must be regarded as essential, and may be employed to great advantage. It will be found, however, exceedingly difficult in most cases, even where there is no depletion from bleeding, to keep up the waning powers, and carry the patient, through this formidable disease, to a favorable termination. Of the utility of warm applications to the surface generally, there can be no question; yet, our main reliance is on internal remedies, as has been already shown: the lost temperature of the body must be restored, the production and diffusion of heat, or caloric, must be internal through the administration of remedies, that will promptly and kindly produce this result.

What are, then, the remedies? We have ventured in this discussion to recommend the internal use of chloroform, and believe it will be found in combination with other prompt and diffusive stimulants, specially adapted to meet this condition. In this recommendation, we feel ourselves fully sustained by the result of various experiments heretofore made, and the recent trials of its use, as an internal remedy in the various stages of the disease.

The earliest record of the use of chloroform in cholera is probably to be found in the London *Lancet* for November, 1848, in which Dr. Hill reports a case of its successful use by inhalation. He placed the patient in bed, covered with warm blankets, and applied friction, stimulant liniments, and heated bags of bran to the surface, and kept the patient under the gentle influence of chloroform, till the more urgent symptoms entirely subsided. At intervals brandy-and-water, and thin arrow-root or milk was given. All other medicines were avoided. Though the urgent symptoms returned at first, as the effects of the chloroform passed off, they were easily controlled by the repetition of the inhalation. By persevering in its use, reaction set in, and the patient became convalescent.

Other cases, afterwards, were treated in the same way, with a similar result. Some, however, required the gentle use of chloroform by inhalation, at intervals, for twenty-four hours; after which, none seems to have been administered. For aught that appears these cases all recovered.

Another very interesting case is related by Mr. Brady, who observes that an elderly lady was seized with slight diarrhœa, which, on the following morning, had become very profuse: excessive vomiting supervened, accompanied by spasms in the calves of the legs, fingers and toes. Under these urgent symptoms, the usual remedy, brandy, was administered without avail; the dejections became incessant, and the spasms increased in intensity, presenting the features of a decided case of malignant cholera. In this condition, the physician was called in haste, as it was believed and affirmed the patient was dying. In describing this case, the physician observes: "On my arrival, I found the patient presenting all the symptoms of malignant Asiatic cholera, in an

CONDITION OF THE BLOOD. 117

advanced stage ; the features collapsed and ghastly ; extremities and tongue cold ; burning sensation in the stomach and œsophagus ; pulsè rapid and scarcely perceptible ; voice diminished to a whisper ; stomach exceedingly irritable, and the dejections from the bowels presenting the characteristic rice-water appearance; and all the voluntary muscles of the body were affected by spasm, so that the patient actually writhed in agony." Ordered the following : ℞. Chloroform ʒj ; Ol. Terebinth. ʒj ; aq. Dist. ʒiij. M. And gave immediately a large tea-spoonful, in a wine-glass, of dilute brandy ; and applied sinapisms to the calves of the legs and abdominal and thoracic surfaces. Thirst was relieved by drinking plentifully of water nearly cold. Though the stomach was irritable, the chloroform was retained, as well as the fluid drank after it, and was followed by no dejection. Half an hour after, two pills were administered, composed according to the following: ℞. Calomel gr. v ; fellis. bov. inspis. gr. x ; Ft. Pil. ij. Half an hour after these were given, vomiting ensued, but soon subsided; the diarrhœa had apparently ceased; the cramps had diminished in frequency and severity. A second dose of chloroform, now one hour after the first, was administered, and soon after this two more of the pills, both of which were retained, and gave decided relief. The pulse rose in power and became slower, the spasms less frequent, and, in an hour after the second dose, the patient was bathed from head to foot in a warm perspiration, and expressed herself comparatively free from all uneasy sensations. The attack had been completely subdued, leaving behind a good deal of pyrexia and debility, from which she rapidly recovered.

Here it is worthy of notice, that in this case, severe as it was, only two doses of the chloroform mixture were administered, each containing about six minims of chloroform and forty of turpentine; the pills would naturally tend to perpetuate rather than relieve the nausea and vomiting, and in one hour after the administration of the

CONDITION OF THE BLOOD. 119

second dose, all the urgent symptoms were assuaged.

In another case, the attending physician reports that, after giving calomel, combined with opium, which was immediately rejected, the following mixture was ordered : ℞. Chloroform ʋj minims ; brandy ʒiij ; water ʒiijss, one-third of which was given immediately, and was thrown up in half an hour ; a second dose was then given, and was retained. The vomiting and diarrhœa ceased ; the spasms became less severe. In two hours after, gave the remaining third part ; and during the next six hours, administered in two doses six minims more of the chloroform, with the most decided benefit, and the patient soon became convalescent. To the extreme tenderness over the region of the epigastrium flannel soaked in spirits of turpentine was applied ; and as no urine was secreted, I am firmly of the opinion that the usual remedies would not have met this case. "I candidly confess," says the physician, " I had no hope of success from its severity ; and, but for a knowledge of Mr. Brady's case, I believe I should have lost my patient."

Dr. Davies reports a case in which he used chloroform fifteen hours after the seizure with relief, but not with success, and observes that, in a number of cases occurring in the hospital, there were 22 cases in which, as severe symptoms came on, the chief remedy was chloroform, administered internally, in doses of from seven to ten minims every hour, half hour, or quarter of an hour, according to the severity of the symptoms. Of these 22 cases, 8 terminated fatally, and 14 recovered.

Again : " Out of 9 cases of cholera, and 13 of the worst cases of diarrhœa occurring in my own practice, and treated with chloroform, *one died*. All these were in the better ranks of life. In some of them, the warm bath (salt water) was used as an auxiliary, and the diet consisted of nothing but cold milk and water, with some carbonate of soda, *ad libitum*. The fatal case was that of a drunkard, who, probably, did not take the remedy. These cases varied in severity, from sickness and diarrhœa, and mild collapse, to sickness, diarrhœa, severe cramps, and great collapse, with almost clear watery evacuations, passing away involuntarily * * Of 14 cases of cholera treated by Mr. Towers, Medical Resident of the Infirmary, many of them under my own observation, *one died*. The fatal case was that of a woman aged 63, who was previously suffering great depression, consequent on extreme destitution."

Again, says Dr. Davies, "It will probably be remembered that, in my second report, I expressed a very favorable opinion of chloroform in this deadly malady. I considered I had strong grounds for so doing, after observing the large proportion of cases which recovered under its administration. From the history of this last visitation in the county prison, however, the fact turns out, that, under some uncertain circumstances, the use of chloroform will not prevent the proportion of deaths being considerable. I have reason to believe that it was, from over-anxiety, given in too frequent doses in some cases, and that it thus rather added to the coma, which is one of the characteristics of the malady.

At the commencement of the outbreak, the doses were repeated every hour, or every two hours, and it is to be noted that the first seven cases *recovered*.

As the cases multiplied, the remedy was given every half hour, and, in some instances, every quarter of an hour; the result was that the next six cases died. Whether these cases had anything in them inherently more fatal, it is difficult to tell. The symptoms at first were about equal, and the differences did not show themselves until towards the end. There was next a recovery of seven cases in succession; in these the remedy was administered less frequently, but subsequently two deaths occurred under the less frequent administration.

The chloroform was administered also by inhalation, in some of the more severe cases of cramps, with the effect of affording relief in every instance. The inhalation was not carried so far as to produce insensibility. Although I am still of the opinion that chloroform properly regulated is the remedy of all others hitherto tried to be depended on, yet it cannot be considered a specific for cholera."

Mr. Steadman reports a very interesting case treated by chloroform. He observes, "The spasms were universal and extremely violent, as if knots were being tied in the bowels, countenance livid and cold, voice feeble, and all medicines rejected. In this condition gave chloroform combined with 'aquæ vitæ' and distilled water. The first dose had a partial but most satisfactory effect. In two hours after, as the symptoms manifested a disposition to return, gave a second dose, which entirely controlled all spasms, vomiting and purging. The patient was ordered cold rice and mucilaginous drinks, and had the chalk mixture with nitric ether prescribed. A dose of oxgall (gr. x) was given in course of the day, which produced the desired effect. In two days the patient was declared convalescent."

The daughter, who had nursed the mother in this case, was seized soon after in a similar manner, except the dejections were more abundant and frequent. The mother having some of the chloroform mixture left, gave it to the daughter without advice or hesitancy, and obtained the same magic results. The first dose was only partial in its effect, but the second completely subdued the disease.

Such are the results of some of the experiments which have been made by the administration of chloroform ; and, so far as appears, the first cases treated by inhalation were severe malignant cholera in the advanced stage, all of which recovered. So, also, those treated by the remedy used internally, combined with a prompt and decided stimulant like the spirits of turpentine, or aquæ vitæ and brandy, recovered. In all these cases the remedy appeared to meet the urgent demand, to remove the impediment to the circulation, to relieve the nausea and vomiting, and purging and cramps, and restore, in a very short time, the general action and normal tone of the system. Still we must admit, that some cases, treated by its internal administration, and also by inhalation, proved, on some accounts not satisfactorily explained, unsuccessful. Were these cases given in detail, it would be much easier to detect the cause of failure, or its questionable use in such cases; but we have only the bare fact that they were thus treated, without the manner or character of the combination, if any were made, being given.

Hence Dr. Davies, under whose direction these cases occurred, remarks, in view of this result, " that *no reliance* could be placed on chloroform alone." The correctness of this opinion cannot be questioned, for the experiments we have cited all show the necessity of a prompt and diffusive stimulant in aid of its action, to render it sufficiently prompt and powerful to meet and overcome the disease in the more rapid and severe cases. Chloroform, properly combined, offers the best hope of relief, and is, without doubt, the most perfectly adapted of any remedy known to the pathology and phenomena of the disease. There is no remedy, when properly combined, so capable of meeting all the indications required as this, and none that can be administered with more certainty of success.

In conclusion, we may, with much propriety, refer again to the pathology suggested by the authors cited above, and inquire whether the action of chloroform as a remedy in these cases be consistent? and whether as such it has that curative influence, or direct controlling power, to arrest, suspend, and cure the disease, so imperiously demanded? We have seen that, according to the opinion generally prevailing, the first impression of the poison is made upon the blood, and through it upon the nerves, especially those which, from their anatomical position, bear the most intimate relation to the blood-vessels. Through this channel the first invasion appears to be made on the ganglionic, the nerves of circulation. These nerves are distributed chiefly to the viscera and blood-vessels, and are at least very early involved and essentially disturbed, for their healthful action depends in no small degree on the aeration or oxydation of the blood. Says an eminent author, " The action of every ganglionic mechanism depends on the existence of certain physical conditions, among which the most prominent and important is the due supply of arterialized blood. If this be stopped but for a moment the nerve mechanism loses its power, or, if diminished, the display of its characteristic phenomena correspondingly declines." Hence the loss of power in these nerves, and their deranged action, the contraction of the capillary and pulmonary arteries, the impaired and impeded circulation and all the phenomena arising therefrom.

Again, the great pneumogastric nerve, which is composed of both motor and sensitive filaments, has a very extensive distribution in the upper part of the abdominal cavity. It supplies the organs of voice and respiration with motor and sensitive fibres, and the pharynx, œsophagus, stomach and heart with motor influence. This very important nerve, through the primary action and deteriorating process of the cholera poison, becomes early involved, and its functions greatly, and, in fatal cases, permanently deranged. The evidence of this disturbance and loss of nerve-power is too obvious to be overlooked or disregarded in the treatment of this disease.

In confirmation of this, we may, with great propriety, adduce the testimony of Dr. Wallis on the loss of nerve-power, and the process through which the result is produced, who observes, that " the phenomena which are exhibited when the deleterious air has been drawn into the lungs are these : the great gastro-pulmonary nerve is either wholly or partially paralyzed, the consequences are the cessation of all its functions either wholly or partially. This great nerve is a nerve of function, and performs the functions of digestion and respiration, and influences all secretions."

Hence it appears the nervous power generally, as before observed, is very early and essentially impaired, and to such an extent that there can be no rational hope of relief, unless some remedial agent can be found that will exercise such a controlling influence and power, as shall be adequate to restore the tone of the nervous system.

Hence, we are forced to the conclusion that the prominent, leading, and most urgent symptoms requiring special attention, are " the Algide" or loss of temperature, the loss of nerve-power in the ganglionic and pneumogastric nerves and their branches, the altered or disorganized condition of the blood, the impaired or obstructed circulation, and the early and direct tendency to congestion. These are the prominent and essential features to be observed in the treatment. They are too intimate, dependent and inseparable, to warrant any attempt to mark the precise order of their development. They are the essential phenomena, proceeding equally and directly together from the primary cause and disease action, and strictly constitute the complex character of the cholera, and exhibit its main, distinguishing features, which must necessarily govern and dictate the maxims of rational practice in the treatment of this disease. The object, then, of first importance is to restore the lost temperature, the caloric already eliminated, and prevent its further depression ; to restore, at the same time, the lost nerve-power to the nerves again ; to arrest the process of disorganization of the blood, and equalize the circulation ; to relieve and suspend the congestion ; and then, according to all the experiments which have been made, the consequent and dependent phenomena of the cramps and the vomiting and the purging will disappear.

Section III.—Different Modes of Treatment.

After speaking of the various expedients resorted to for the cure of cholera, says Dr. Watson : "I believe that each, in some cases, did good, or *seemed* to do so ; but I cannot doubt that some of them did sometimes do harm. I had not more than six severe cases under my own charge, and I congratulated myself that the mortality among them was not greater than the average mortality. Three died, and three, I will not say were cured, but recovered, * * * under large and repeated doses of calomel. Yet, as I said before, I do not venture to affirm that the calomel cured them." It seems that Dr. Latham commenced the treatment and Dr. Watson followed it up, repeating the half-drachm doses of calomel many times, as the patients seemed to rally after its administration. Again, he observes : " It was remarked of those who recovered, that some got well rapidly and at once, while others fell into a state of continued fever, which frequently proved fatal, some time after the violent and peculiar symptoms ceased. Some, after the vomiting and purging and cramps had departed, died comatose—*over-drugged*—sometimes, it is to be feared, by opium. The rude discipline to which they were subjected might account for some of the cases of fever." * * * " Never, certainly, was the artillery of medicine more vigorously plied, never were her troops, regular and volunteer, more meritoriously active. To many patients, no doubt, this busy interference made all the difference between life and death. But if the balance could be fairly struck and the exact truth ascertained, I question whether we should find that the aggregate mortality from cholera in this country was any way disturbed by our craft."

In a report by the acting physician to the Bellevue Hospital, made to the "Special Medical Council," August 2d, 1832, while the Epidemic Cholera was still prevailing there and in the city, the physician says: "The treatment I have divided into two kinds—the pathological and the mixed. The first having been determined on, after the careful examination of twenty-three persons dead of cholera; since then, ten more have been examined, which serve to confirm the conclusions first formed.

PATHOLOGICAL TREATMENT—*First Stage:* This consisted in the administration of blue pill and opium with absolute diet. If pain was present, leeches to the epigastrium and arms, and when these could not be procured, cups to the epigastrium. This plan never failed to arrest the disease in the hands of those who diligently pursued it, where the mucous membrane of the gastro-intestinal canal was not previously diseased.

Second Stage.—First, Blood-letting; second, diligent frictions with the ointment alluded to above, when persons could be procured to perform the duty; third, ice to allay the thirst; fourth, small doses of brandy and laudanum, if the vomiting continues; fifth, cups to the epigastrium, if there was pain and the brandy omitted.

Third Stage.—First, ice to allay the thirst, which is now, indeed, unquenchable; second, external heat; third, a continuation of the frictions; fourth, no opium, and, frequently, no brandy, especially among the children.

MIXED TREATMENT—*First Stage.*—Besides the above treatment, calomel and Dover's powders was a very frequent prescription ; also scruple doses of calomel, and calomel and opium in small doses, and all with success. Nevertheless, I believe they occasionally did harm.

Second Stage.—First, blood-letting less frequent than above ; second, calomel and Dover's powders continued ; third, calomel and opium ; fourth, calomel, capsicum and opium ; fifth, soda powders ; sixth, scruple doses of calomel every half hour ; seventh, ice.

Third Stage.—Calomel and Dover's powder ; calomel and opium ; calomel, capsicum and opium ; carbonate of ammonia and capsicum ; scruple doses of calomel every half hour. External heat in various ways ; ice, etc. Severe shocks of electricity along the course of the muscles to allay the cramps ; also, the burning of alcohol on the skin. The first was the practice of Dr. Devan, the second, that of Dr. Gardner, and both lay claims to having been the first to use these means."

The ointment alluded to above is composed of mercurial ointment, one pound, camphor finely pulverized, seven ounces, and the same quantity of capsicum. With this, the patient was rubbed briskly from head to foot and repeated at short intervals. The result was, that mercury generally showed its specific effects upon the gums in from five to ten hours from the commencement of reaction. The success of this external application of mercury, conjoined with its internal administration and frequent bloodletting, may be learned from the cholera statistics of this and other institutions.

Dr. Pereira employed sixty-grain doses of calomel, it is said, with success, and Dr. Barton of New Orleans, in 1849, gave in ten cases from 120 to 150 and even to 180 grains of calomel at a dose, and, in one case, gave 220 grains, intending, it is said, to have weight sufficient to keep it down. This brave and heroic practice did not afford relief in a single instance; the cramps, and vomiting, and purging continued, and a few hours closed the scene—all died.

The treatment recommended in the American Practice of Medicine, by Dr. W. Beach, which was fully tested by the author himself while in discharge of his official duties as physician of the Tenth Ward, city of New York, during the prevalence of cholera in 1832, is worthy of consideration on account of its simplicity, its great efficiency and wonderful success. "Among all the medicines," says the author, "ever given or proposed in the incipient or premonitory stage, none will be found so efficacious as our neutralizing mixture, made of genuine materials and given very strong. Occasionally, it may be proper to add fifteen or twenty drops of laudanum; this, however, is very seldom necessary. A vast number of medicines are recommended in this stage of cholera, but there are none, I am convinced, so efficacious as the above."

In the second, or confirmed stage, the same medicine was continued in larger and more frequent doses, with hot fomentations to the abdomen, stimulating lotions, sinapisms and injections. The cholera drops were also administered, composed according to the following formula:

℞. Tincture of Capsicum,
   Tincture of Opium,
   Spirits of Camphor,
   Essence of Peppermint.
                    Equal parts—mix.

Give a tea-spoonful every hour or half hour, according to the severity of the symptoms.

In the third, or collapsed stage, he directed a tea-spoonful of pulverized black pepper to be mixed and given in a tumblerful of hot gin-sling; also, the same to be prepared and applied hot to the bowels and extremities. Also, to two tea-spoonfuls of either pulverized red or black pepper, pour on a sufficient quantity of hot water, let it stand till nearly cold; strain and inject the whole up the bowel. This would often arouse the patient in the collapsed stage when there was little or no hope of recovery. Such are in brief the remedies which were used so successfully in the Tenth Ward of this city, in 1832. Here it will be noticed that the general principal evolved in this treatment consists in its prompt and diffusive stimulant, its antispasmodic and corrective power so combined as to act gently and kindly, yet promptly and successfully, as the records show, to which we shall refer in the sequel.

Another mode worthy of a passing notice is one analogous to this, adopted and recommended by the eminent Dr. G. S. Hawthorne, of Liverpool, England, who observes: "Of the medicinal remedies, the chief is opium. This, I have explained, should be given in combination with medicines of a cordial, stimulating and antispasmodic character, of which the most efficient are camphor, capsicum, ether and aromatic spirits of ammonia. The following formulæ present the combination of the medicines which I would prefer:

℞. Powdered Opium, . . . gr. xij.
    Camphor, ... gr. xxx.
    Capsicum, ... gr. ix.

Spirits of wine and conserve of roses Q. S.—mix—divide into twelve pills. Each of these pills, it will be observed, contains one grain of powdered opium. These are accompanied with the following :

℞. Chloric Æther,
    Aromatic Spirits Amomnia,
    Camphorated Spirits,
    Tincture of Capsicum.
               Of each, one drachm.
    Cinnamon water, two ounces—mix.

" Cholera," observes Dr. H., " presents itself in four distinct degrees of malignity. All the modifications of the disease require to be treated on the same principles, the only difference being that, in the detail, the milder forms require less powerful doses of the medicines. The mode of treating the most malignant form of the disease, will serve as a model on which all the others are to be treated. This most malignant form has, by all writers on the subject hitherto, been pronounced incurable. They say it never was cured in a single instance, and never can be cured by the power of medicine. I shall, however, point out a mode of treating it which will prove itself infallibly successful where my directions are followed with sufficient promptness, boldness and skill." In detailing the mode of procedure, the doctor observes: "Place the patient immediately in the horizontal posture in bed, and give him on the instant, as this is an extreme case, ten of the antispasmodic pills, and two ounces of the antispasmodic mixture, and wash the whole down with a glass of undiluted brandy or whisky, flavored strongly with cloves, essence of ginger, or some such warm aromatic spice. In the mean time, have him covered with an additional blanket, and let the usual means of communicating heat, such as jars or bottles of hot water, bags of hot salt or sand, hot bricks, or whatever can be most readily procured, be applied without delay to the feet and different parts of the body, so as to restore the temperature and produce perspiration as quickly as possible. As soon as the perspiration has begun to flow freely, superadded to the medicines and cordials already administered, a glass of brandy-punch should be given, the punch to be made strong and to be swallowed hot as possible. After this, no drink should be given until the perspiration has flowed freely for a few minutes. The stomach will then retain it, and the patient should be indulged freely with copious draughts of rennet whey, warm toast-water, flavored with some agreeable spice, mint, or balm-tea, or any such mild beverage. The necessity of attending to this is most important. When the discharges from the bowels cease, and when the pulse becomes full and bounding, the body is covered with a copious, warm perspiration, which will not fail to be the case under such treatment; the danger is over. The perspiration, if the patient can bear it, should be kept up for twelve hours, and may, with advantage, be continued moderately even longer. Its duration, however, must be regulated according to the strength of the patient and the state of the pulse. After the first four or six hours, more heat need not be applied than is perfectly agreeable to the feelings of the patient. It is remarkable how sud-deny the precordial oppression, etc., are relieved on the breaking out of a free perspiration, and, what is of greater importance still, the vomiting, where it exists, immediately ceases." In short, all the urgent symptoms soon subside, and the patient becomes convalescent.

Such is Dr. Hawthorne's treatment, which is affirmed to have been invariably successful. It is based on the same general principle as the preceding—a prompt and diffusive stimulant. Here we might ask, What constitutes the chief reliance in the formulæ? Was it the opium that so promptly met and arrested the disease? or the combination of the other powerful stimulants with which it was united? Dr. H. places his main reliance on this drug, and yet affirms that it produced no narcotism or other sensible effect whatever, except as a diaphoretic, and even in this its influence may be questioned. The prognosis becomes favorable from the fact of a sudden rise in the temperature of the body, for the icy-coldness disappeared, the heat of the surface returned, the circulation was equalized and a profuse perspiration set in, and, as these conditions appeared, the urgent symptoms subsided. Not the excessive doses of opium, but the remedies in combination as a whole, produced by its prompt stimulating power these results, and the patient was thus relieved.

Mr. Forward, while superintending some of the public works in the State of Kentucky, in 1832, had in his employ more than two hundred laborers, among whom the Cholera Epidemic of that year appeared about a week before its irruption in Louisville. The first case was that of a young, sober, industrious white laborer, who was at the time vigorous and apparently healthy. It was a sudden and severe, case and occurred about eleven o'clock at night. The physicians who usually attended these men were at a distance, and could not be obtained without considerable delay. Under these circumstances, Mr. Forward, after visiting the patient, becoming acquainted with the symptoms, and believing it a genuine case of cholera, commenced treatment at once, fearing, as he states, the patient could not live till a physician could be obtained. It was, indeed, a desperate case; violent spasms, with constant vomiting and severe purging, attended with that livid appearance and peculiar coldness so characteristic of the disease. "Of the treatment," says Mr. Forward, "I gave him first a quick, stimulating emetic prepared from the lobelia seed, which checked the vomiting and purging, but had little effect upon the spasms. I then applied the steam bath, having his feet and legs at the same time immersed in water as warm as he could bear, which was made strong with salt and wood ashes. I then sweetened a tumbler of warm water and put into it a teaspoonful of "number six," and about the fourth part of a tea-spoonful of Cayenne pepper, and gave him one-third of it when I commenced sweating him, and the balance at intervals while he was sweating. By the time he had been sweated ten minutes, he was free from spasms and pain, but I continued the sweating ten or fifteen minutes longer, then wiped dry, after which the patient laid down and went to sleep—being thus relieved and cured."

Another case of a colored man who was strictly temperate and healthful occurred an hour or two later the same night. His attack, too, was sudden, and still more severe; cramps very violent, vomiting and purging equally as severe, though he had not been awakened from his slumbers more than fifteen minutes. This case was treated the same as the former, with the emetic, sweating, and when the sweating had subsided, administered a table-spoonful of spirits of turpentine, which relieved him entirely, and he soon went to sleep. The next morning both were comfortable, and went to work and remained well. During the prevalence of cholera at that time, Mr. Forward had thirteen cases in his own family, and, on one day when the epidemic was at its height, seven cases among the laborers. All these and many others that occurred were treated in the same manner, with the same undeviating success. Not a single instance of death from cholera in his own family, or among the hands on the road. When the epidemic cholera reappeared in 1835, the same course of treatment was pursued, with the same uniform success. Such results, considering the malignant character of the disease, are truly astonishing. Whatever may be said of the general principle of practice in these cases, its success must be admitted as equaling, if not surpassing, the treatment of any equal number of cases on record. Though conducted by an unpretending and unprofessional gentleman, yet, out of the whole number attacked during the continuance of the epidemic, not a single case was lost.

In a report of a case of cholera treated successfully by rectified oil of turpentine, administered internally as a specific, by Richard Brown, Esq., Surgeon, Cobham, Surrey, November, 1848, it is stated that the patient, " aged fourteen, having suffered from severe bowel complaint, presented all the symptoms of cholera in the stage of collapse. The bowels acted incessantly, and anything taken into the stomach was immediately rejected; the pain around the umbilicus was intense, attended with severe cramps of the legs; the pulse exceedingly small, and scarcely perceptible; tongue coated in the centre, and flabby; the surface of the body much below the natural standard; the countenance of a blue cast, and expressive of the greatest anxiety. So decided, indeed, was the symptom that the case was considered almost without hope." "But I had determined," says the physician " to treat the first case of cholera that occurred in my practice with rectified oil of turpentine, given internally, the active principle of which, camphogen, possesses stimulating, diuretic, diaphoretic, sedative, antispasmodic, antiputrescent properties. I administered immediately one drachm of it combined with mucilage and aromatics, directing it to be repeated every two hours, the patient to be kept warm and to take meal broth with excess of salt."

Now mark the result of this simple, uncombined remedy. In the evening of the same day all the urgent symptoms were assuaged, the purging and vomiting had ceased, the pulse was raised, the surface of the body had become warm and moist with perspiration, the pain around the umbilicus diminished, and the cramps less violent, but the countenance still bore the appearance of great anxiety. Such were the immediate results of the administration of this remedy, which appear, from the subsequent history of the case, to have been permanent and unattended with any constitutional derangement, or other serious and unpleasant effect. On the morning of the next day the patient was steadily improving; much of the anxiety of countenance had vanished, but the pain in the belly and cramps of the legs still remained, though much relieved. On the second morning after the attack the patient was very much better; no pain in the belly, and does not feel sick from the turpentine. On the third morning the patient was up, and, though exceedingly weak, there was no trace of any alarming symptom re-

DIFFERENT MODES OF TREATMENT. 149

maining. The bowels had moved from the effects of a previous dose of calomel (two grains) given the next morning after the attack, and the evacuation was much more healthful. A mild tonic and alterative plan of treatment was all that was necessary to restore the patient to her usual health, and she is now well. The remedy was given at first every two hours, then every four, and lastly every six hours. This treatment commenced on the 26th and terminated on the morning of the 29th. Its duration about sixty hours, when the patient is declared convalescent and comparatively well. Here we might ask, What experiment with any single remedy has been more important and satisfactory in indicating and directly pointing out a general principle of practice for the successful treatment and cure of this formidable disease? We say single remedy, for it is doubtful whether the two grains of calomel exercised any curative influence whatever, or in any way varied the result. It is, therefore, to the use of the rectified oil of turpentine that the favorable termination and cure of the disease is to be attributed.

There is another mode of practice which has been exhibited to some extent in almost every part of the world, claiming to be more efficacious and successful than any other in the cure of epidemic cholera. It is the general principle which is the great and important consideration with which we are concerned in presenting it among the various modes adopted for the cure of this disease. This is found clearly defined and ably presented by Dr. Joslin in his lecture on cholera, in which, after exhibiting the views and doctrines governing the practice, and contrasting its results with those of other modes, he observes, in relation to the treatment of cholera in its early stages, that "whatever may be the form of attack, give one drop of the tincture of camphor dropped on a lump of sugar, and then dissolved in a table-spoonful of cold water. Repeat this every five minutes until there is a decided mitigation of the symptoms. This will usually be after five or six doses.

DIFFERENT MODES OF TREATMENT. 151

If the disease be taken in time, ten or twelve doses are ordinarily sufficient. There is abundant testimony of the efficacy of this camphor treatment from all parts of Europe." Again, speaking of the first variety, in which the most prominent symptom is diarrhœa, the Dr. observes, " If camphor does not soon give relief, we are to resort to phosphorus, or to phosphoric acid. Dr. Quinn has employed both with equal success. Phosphoric acid is to be preferred when there is a gluey matter on the tongue. In some cases, veratrum, chamomilla, mercurius, or secale may be indicated. However, phosphorus and phosphoric acid rarely fail to cure ; and some high authorities are in favor of giving one of them at first, in preference to the administration of camphor in this form of cholera."

Again, in the second variety, cholera gastrica, Dr. Joslin observes, that " the remedies are generally ipecacuanha or veratrum, sometimes nux vomica. Camphor is to be given at the outset. Put two or three globules of the third of ipecac. in a little sugar of milk and place them on the tongue. This may be repeated, if necessary, in half an hour, an hour, or an hour and a half. But if the disease is not checked, give veratrum or other medicines according to the different indications." Again, in the third variety, cholera spasmodica, " the remedies are camphor, cuprum metalicum, and veratrum. If camphor has not relieved, give cuprum, and repeat it many times, at intervals of half an hour or an hour, if its salutary effect is not manifested. If necessary, then give veratrum in repeated doses, or other medicines, according to the different indications." In the fourth variety, cholera sicca, " there is no diarrhœa or vomiting; there is sudden prostration of the vital powers," etc. " The first remedy, as in other varieties, is camphor. If the patient is cold, blue, pulseless, that is, collapsed, carbo vegetabilis ; some recommend hydrocyanic acid." In the fifth variety, cholera acuta, veratrum is named as the main remedy.

Such is, in brief, the treatment so highly

extolled and recommended by some in the cure of cholera. It is, in substance, the same as was originally suggested when the disease first appeared in Europe, nearly half a century ago, and will probably continue unchanged for generations to come. Of its general principle and its adaptation to the pathology of the disease we shall speak more at length in the sequel.

After referring to the pathology of the epidemic cholera, showing its strong analogy to congestive fever, from the fact that in both diseases the blood recedes from the surface, and collects upon the internal organs, inducing a state of congestion, and showing the necessity of adopting prompt and efficient means to promote reaction, Dr. Massie observes, "I am not so bigoted, or so wedded to any system of medicine, as to be its champion to the exclusion of others. I consider I have a perfect right to investigate all of the different systems, and avail myself of any information which I may deem important and true, and I will premise by saying that the treatment I now adopt for cholera has been attended with more success than when I treated it under a different system."

"If I am called at an early period of the disease, even when there is nausea, vomiting, and diarrhœa, I commence the treatment by giving equal parts of rhubarb root pulverized, saleratus, and peppermint plant powdered; one pint of boiling water being added to half an ounce of this compound. After simmering it for half an hour, sweeten with loaf sugar and strain, and, when nearly cold, two or three table-spoonsful of good French brandy should be added. Give two table-spoonsful of this, taken warm, in connection with the following preparation, viz.: ℞. Pulverized cinnamon, cloves, and gum guaiacum, each one ounce, good brandy one quart, given in two tea-spoonsful to a table-spoonful every fifteen or twenty minutes to an adult."

"The patient should be well covered with warm clothing, and bottles of hot water, bricks and stones placed around his body. This course is almost sure to be followed by a moderate moisture of the skin, which should be kept up for eight or ten hours; to do which, I give ptisans of catnip or spearmint, and apply hot tincture of Cayenne by flannel cloths over the abdomen; if this fails to keep up the perspiration, I administer the following: ℞. Camphor, grs. x.; Ipecac., grs. v.; Opium. grs. ijss; Supercarbonate of soda, ℈ij. Mix, and divide into two, three, or more powders; give one every hour, or oftener."

"In very urgent cases, I have used tincture of camphor, ℥iv; essence of peppermint, ℥iv; syrup of ginger, ℥ss; tincture of Cayenne, ʒj. A table-spoonful, from one to four in an hour. I have given the saturated tincture of prickly ash, with the compound tincture of guaiacum, with good effect, in doses from a tea-spoonful to a table-spoonful every fifteen or twenty minutes. When there is excessive irritability of the stomach, the following injection should be given after every discharge: ℞. Saturated tincture of prickly ash, ℥ss; water, ℥j; tincture opii, ʒss. Mix." Such are the views of

Dr. Massie, as presented in his Treatise on the Eclectic Southern Practice of Medicine. They are confirmatory of the observations and experience of many other eminent practitioners, and strictly accord with his views of the pathology and essential phenomena of the disease.

We find in a very valuable work, entitled the Eclectic Practice of Medicine, published at Cincinnati by Professors Powel and Newton, a full account of the mode of practice generally adopted and pursued by the great body of physicians in the West, the substance of which we are induced here to present, preserving, as far as practicable, the language of the authors. For our inquiries are, What are the modes of practice? and what modes, if any, are consistent with the pathology and the essential phenomena of the disease? Each mode, however prominent or however obscure, is entitled to a fair representation in our inquiries, and should be held responsible for its deviations from the strict and generally received principles of science, and the consequences arising from any such deviations, or departure therefrom.

"When called upon," say these eminent professors, "to treat a patient in the early stage of the disease, he should at once be placed in a recumbent position, and everything should be avoided which will have a tendency to disturb the mind, as well as the stomach and bowels. In the greater part of cases in this early stage, the administration of the compound pills of camphor, made according to the following formula, is sufficient to prevent a further development of the disease:

℞. Camphor, ⎫
   Opium,   ⎬ āā., gr. xxxv.
   Kino,   ⎭
   Capsicum, ... gr. v.
   Conserve of roses, Q. S.—Mix.

Divide into thirty pills, and give one after each discharge from the bowels, or oftener, if the urgency of the case requires it. Occasionally, however, there may be applied a large sinapism over the whole abdomen with advantage. Greenhow's aromatized brandy,* the aromatic tincture of guaiacum,* may sometimes be beneficially alternated with this pill. Should there be an overloaded condition of the alimentary canal, the fluid extract of rhubarb and potassa,* three parts, with saturated tincture of prickly-ash berries, one part, may be administered in table-spoonful doses every hour, and continued until the bowels are properly evacuated, after which the above astringents may be given; but where the diarrhœa is excessive, it would be imprudent to wait for catharsis, as the discharge should be checked as speedily as possible.

In the second stage, when nausea, vomiting, and cramps are present, more active means should be pursued. To overcome the nausea or vomiting, the preparation of Dr. O. E. Newton, termed in the American Dispensatory compound mixture of camphor,* may be used with excellent effect; it is prepared as follows :

* See American Dispensatory.

℞. Camphor water,
   Peppermint water, } āā., f ℥j.
   Spearmint water,
   Paregoric, ... f ℨij.
                              Mix.

From a tea-spoonful to a table-spoonful may be given every five or ten minutes; and in cases where this does not act sufficiently prompt, the following may be administered:

℞. Common salt, . . . ℨj.
   Black pepper, . . . ℨj.
   Vinegar, ... f ℨv.
   Hot water, ... f ℥iv.
                              Mix.

Of this a table-spoonful may be given every ten or twenty minutes, and continued until the nausea ceases.

To remove the cramps, hot bricks, or bottles of hot water, etc., should be kept applied to the feet, legs and arms, and cloths wet in water as hot as can be borne, must be applied over the abdomen and changed every few minutes; this should be perseveringly pursued until relief is obtained. Sometimes advantage will ensue from stimulant applications along the whole length of the spine. Cramps of the muscles of the limbs may be overcome by bathing with the compound cajeput mixture,* either alone or in combination with chloroform, and applying friction at the same time. This course usually checks the further progress of the disease, and the patient is saved; however, should it fail and the stage of collapse come on, in addition to the above treatment energetically pursued, the patient should be enveloped in blankets, wet with water as hot as can be borne, which should be renewed every ten or twenty minutes, and stimulants may likewise be given; the saturated tincture of prickly-ash berries will here be found beneficial, both by mouth and enema."

Dr. Morrow observes, that " to fulfill the most prominent indication, the production of an equilibrium in the circulation, and excitability, the compound tincture of guai-

* See American Dispensatory.

ac* may be given. This is prepared by adding gum guaiacum, cinnamon and cloves —each, one ounce to a quart of best brandy, and is administered in tea-spoonful does in hot, sweetened water and brandy, every fifteen or twenty minutes till relief is obtained. As a general remedy, its exhibition is most salutary. In some cases where excessive nausea is the most prominent symptom, it may be advisable to administer an emetic to relieve the gastric irritability, to equalize the circulation and check the spasms. For this purpose, the acetous tincture of lobelia and sanguinaria,* with the addition of one-third spirituous tincture of aralia spinosa,* is preferred. This is given in doses from a tea-spoonful to a table-spoonful every ten minutes in warm catnip-tea, sweetened. In very urgent cases, it may be given in larger doses and frequently repeated.

In most cases, the saturated tincture of xanthoxylum fraxinifolium bac. may be used with great advantage. It is a reliable, excellent and prompt remedy. When given

* See American Dispensatory.

in the early stages, it will frequently relieve in from ten to twenty minutes. In combination with the fluid extract of rhubarb and potassa,* it has generally proved very prompt and efficient. In cases of partial collapse, when the patient is suffering from severe cramps, Hunn's Antispasmodic Mixture * is an excellent remedy. In cases of violent spasms, it has been administered every ten minutes in doses of from one to two tea-spoonsful in hot brandy-and-water sweetened, with great advantage, and it is peculiarly applicable in such cases where there is not too great irritability of the stomach. In many cases, camphor is very beneficially prepared, by adding one drachm of camphorated spirits to a half-pint of cold water and the mixture given in teaspoonful doses every three or four minutes.

Dr. King states that in the early stage he has used very extensively the following preparation:

℞. Ox Gall, ... ℥j.

* See American Dispensatory.

Capsicum, ... ⎫
Gum Guaiac, . . . ⎬ āā., ℈iv.
Leptandrin, . . . ℨiv.—Mix.

This was given in doses of one grain, and repeated two or three times a day. He had also succeeded in some cases with a mixture composed as follows:

℞. Sulphur Sub., . . grs. iv.
Gum Guaiac, . . grs. ij.
Charcoal, ... grs. ij.
Camphor, . . . gr. j.
Opium, ... grs. ss.—Mix.

Dose, one to ten grains, repeated every ten minutes until relief is obtained. In some cases, however, this compound did not appear to exercise any beneficial influence. In cases of excessive irritability of the stomach, oat-meal cake coffee was given, for the purpose of allaying its irritability, with admirable effect. The saturated tincture of prickly-ash berries,[*] combined with tincture of opium, was used in some cases as an injection, with very good effect.

Dr. R. S. Newton observes that he had

[*] See American Dispensatory.

also used a preparation composed of equal parts tannin, capsicum, camphor and kino, with considerable success, to be given in doses of four grains, and repeated at short intervals until the discharges were checked.

He considered the saturated tincture of xanthoxylum fraxinifolium bac.* the most valuable of all the remedies for the cholera which he had tested. When the stomach would not retain it, he gave it as an injection. It had a peculiar influence on the system, and having taken the remedy, he could speak from experience of its effects. When given as an injection, the effect produced was almost instantaneous; the sensation was as if he had received an electric shock; its use was very soon followed by a copious perspiration. He had more confidence in this than any other one remedy with which he was acquainted.

Dr. Wright observes that he had also used the neutralizing extract, saturated tinc. xanthox. fraxi. bac., and the compound tincture of guaiac.* He had succeed-

* See American Dispensatory.

ed best with a mixture of equal parts tincture of prickly-ash berries and neutralizing extract.*

He had always found it necessary to attend strictly to the surface. The best external application he found was equal parts of capsicum, salt and mustard.

Dr. Chase states that, "in the early period of the disease, he had used the leptandrin, combined with neutralizing extract,* very successfully. He thinks opium can be dispensed with in the treatment of cholera altogether. In typhoid cases, he pursued an entirely different course, and remarked that many cholera cases presented symptoms similar to those described in Wood's Practice, as belonging to pernicious fever, which must be treated according to their peculiar character."

Such, it is said, is the more general and successful practice in the Mississippi Valley, where the disease has several times prevailed in its most malignant form. For its curative efficiency much is claimed. Its utility, however, must be measured, as in all

* See American Dispensatory.

other cases, by the unerring rule, the actual results sustained by incontrovertible facts. The nearer any mode of practice accords with the general principle of pathology, the greater must necessarily be its success, for it is not in this disease, or in any other, that the bold, energetic and heroic practice, which is inconsistent and incompatible with this principle, cures, however extensively adopted and rigidly pursued. For this principle must direct and govern the practice, or else it becomes necessarily experimental or empirical, and must be inevitably attended with the most lamentable results.

Section IV.—Statistics—Percentage of Loss—Variable Results—Their Cause.

The results of the different modes of practice which we have briefly noticed will aid materially our effort to discover and establish some general principle for the successful treatment and cure of cholera. For all modes, whatever be their merits or demerits, are supposed to be founded on the pathology of the disease. To treat any disease successfully, its pathology must be observed, and so applied in the arrangement and adoption of a mode of practice as to secure not only entire harmony, but a complete and perfect adaptation of the treatment to its pathological character. The nearer any mode approaches to an exact conformity to this principle the greater will be its success. The neglect to conform, in the treatment of the epidemic cholera, to this acknowledged and universal law, has, no doubt, been the prolific cause of the sacrifice of thousands of valuable lives. For this principle is the key to unlock the mystery of disease, unfold the process of dseased action, and, as an unfailing and definite rule, must govern all correct theories as well as all rational practice of medicine, under whatever name it may be conducted. All practice, then, deviating from, opposed, or contrary to, this principle must be purely empirical, and unworthy the confidence of an intelligent community. Hence we may refer to statistics rather than argument on the subject, to ascertain how far and to what extent each of the different modes of practice conform to the general principle ; and on the other hand, to show what modes may be at fault, being deficient in the application of science, opposed to the established laws of practice, and contrary to observation and experience, and therefore utterly and hopelessly empirical.

The statistics collected from the most reliable sources, and here presented, may be regarded as a fair representation of the general average of loss by the different modes of practice. In a report now before us, it is stated, " The average proportion of deaths in Paris from cholera, treated under the allopathic practice, was 49 per cent. ; while that under the homœopathic was only $7\frac{1}{4}$ per cent." " In Vienna, (Aus.,) under the former, the deaths are reported at 31 per cent. ; while under the latter it was only 8 per cent. In Bordeaux, death occurred under allopathic treatment at the rate of 67 per cent., and under homœopathic, 17 per cent. only. The general average in the places last mentioned will stand thus: Allopathic, 49 per cent.; homœopathic, $10\frac{1}{4}$ per cent." The record of mortality in twenty-one hospitals in Europe shows the average deaths under allopathic treatment to be $65\frac{1}{8}$ per cent., while in ten hospitals where the cholera patients were under homœopathic treatment, the average deaths from that disease was $11\frac{3}{4}$ only. In a report "published by the authorities of Pischnowitz (in Prussia), it will be seen that 680 cases were treated as follows: 278 treated homœopathically, of which 27 died; 381 treated allopathically, of which 102 died."

In St. Louis, during the prevalence of cholera in 1849, the number treated by three homœopathic doctors, to July 13th, was 1,567, of which 51 died—a loss of $3\frac{1}{4}$ per cent.

In Cincinnati, during the month of May, there were treated by the eclectic physicians 330 cases of cholera and 198 cases of cholerine, of which only five died.

In the same city, during the same time,

there were treated by the allopathic physicians 432 cases of cholera, of which 116 died.

Again, during the month of June there were treated by the eclectic physicians, when the disease had reached its maximum intensity, and many of the patients being reached by the physicians only in the collapsed stage, 764 cases of cholera, with a large number of choleroid diseases not fully reported. During this month, the mortality with all physicians was necessarily greater than either in the preceding or subsequent month. Including then the month of May, the aggregate to July 1st is 1,094 cases, with a loss of only 36, which is considerably less than four per cent. (being 3.28); while the mortality of the old school cholera practice being 26 per cent. in May, must have risen to at least 50 per cent. in June, when the ratio of mortality was more than doubled with all physicians. The *Western Lancet* for July, 1849, issued while the cholera was still raging, and speaking in behalf of the allopathic physicians, observes, "that of the cases of true cholera, with rice-water discharges, at least one-half the cases in this city, as everywhere else, proved fatal." This confession of the *Lancet*, edited by a thoroughgoing allopathic physician, advocating the interests of that school, must be regarded below rather than above the actual allopathic loss. Now, admitting the *Lancet's* correctness, and taking into account the aggregate loss of only 36 by the eclectic physicians in treating 1,094 cases of " true cholera," we ask what must have been the loss by the allopathic school of practice to have brought the average percentage of all schools up to 50 per cent., as affirmed by the *Western Lancet?* If the cholera hospitals be included in exhibiting the results of the different modes of practice, it will appear from the reports that the total number of deaths, compared to the admissions, was, under the eclectic treatment, $23\frac{1}{3}$ per cent.; under the allopathic treatment, 60 per cent. This percentage is confined exclusively to the three cholera hospitals reported.

In the report of 1832, by Dr. Atkins, it appears "that the total number of cases" of cholera in this city, New York, "including those in the hospitals, as well as those reported to the Board of Health, had been 5,835 on the 1st of September. The total number of deaths by cholera to September 1st was 2,996." More than one-half died. "Dr. Buell reports the success," says Professor Clark, "of sixty-grain doses of calomel in one of the New York hospitals, as 93 deaths in 100 cases;" very remarkable success! the largest mortality in the city.

As like causes produce like effects, we need not be surprised at this high rate of mortality, for, says Professor Aikin, "taking the whole number attacked, it is said that the number of deaths in Astrakan were *as one to three;* in that of Mishni Novogorod, *as one to two;* in Moscow and Kasan, *as three to five;* and in Penza, in the country of the Don Cossacks, *as two to three.* In the summer of 1831 the mortality at Riga, St. Petersburg, Mittan, Limburg, and Brody, according to the *Berlin Gazette*, was *about one-half*, while at Dantzic, Elbing, and Posen it was *about two-thirds* of the whole number attacked. The period of the epidemic, however, greatly influenced the mortality; for on the first onset, *nine-tenths* of all those attacked perished, then *seven-eighths;* and the proportion of deaths forms a gradually decreasing series of *five-sixths*, *three-fourths*, *one-half*, *one-third*, till, towards the close, a large proportion of those attacked recovered. The uniformity of this law in every country affected with cholera, whether Europe, America, India, or China, is extremely remarkable." This high rate of mortality is truly and peculiarly illustrative of the inadaptation of the general mode of the so-called regular practice to the pathology of the disease. This, no doubt, is the main cause of its failure, and justly exposes it to the unenviable distinction of being empirical.

The practice of Dr. Beach, the physician of the Tenth Ward of this city, during the prevalence of the cholera in 1832, embraced about one thousand cases, of which only a small percentage was lost. One of his associates, Dr. Hopkins, reported 157 cases, of which only 6 died, being less than 4 per cent., which probably is not much below the general average of the other districts in that ward at that time.

Mr. Forward, an unprofessional gentleman of Kentucky, treated a large number of cases, during the prevalence of the disease among his employees, numbering over two hundred, without the occurrence of a single death. Another instance similar in principle is that of Dr. Browne, who reports a case treated by rectified oil of turpentine, with the most satisfactory and happy result. So, too, the late Dr. Sharp, of Paris, Ky., adopted a similar principle of practice, and became, thereby, eminently distinguished for the cure of cholera; his percentage of loss being very small indeed.

We might extend these statistics and references, and quote from the reports of many other distinguished physicians who have been very successful in the treatment of this disease; but these are sufficient for the purpose of directing our inquiries as to the utility and success of different modes of practice. It is immensely important to ascertain, if practicable, the general principle which has been most successful in the treatment of this disease, before it shall again make its appearance among us as a prevailing and fatal epidemic; especially when we realize and duly appreciate its vast mortality, as represented in the report now before us, that prior to its recent irruption and prevalence in India and Europe, nearly fifty millions of the earth's inhabitants have been swept away by this terrible scourge alone.

This estimate may, however, appear excessive and unworthy belief. Yet the general average for the forty-three years included is only a little over one million per annum, truly a vast number to be carried off by the prevalence of one disease alone. But, if we reduce this estimate within more reasonable limits, and take only two-fifths of it, or twenty millions, as an approximation to the truth, it would still be appalling, and imperatively demand, on account of the vast interests involved, the most rigid and thorough investigation as to both the direct and indirect cause of this vast sacrifice. It will also furnish us a sufficient apology for attempting a brief review and critical examination of the principles involved in the different modes of practice noticed above, in order to ascertain any failures or errors that may have, in some degree, operated as the indirect cause, in procuring this immense loss of life. All must admit that there are, in respect to the treatment of the cholera, great and palpable failures and errors which, though they have continued for nearly half a century, and have been sanctioned by high authority, as well as by long usage, ought nevertheless to be fully shown and exposed, so that they may henceforth be avoided. In our examination, there is but one rule to be observed, and one criterion of ultimate appeal by which to try each and every principle on which any mode of practice may be conducted. This universal and acknowledged rule is Pathology, the science which unfolds and exhibits the nature and character of disease, and "dictates the maxims of rational practice." It is the foundation and only base of rational medicine, which proceeds on the assumption that the nature and character of disease is fully known and appreciated. This knowledge is not only rational, but indispensable, in order to understand and apply the principles which ought to govern in the medication and cure of disease.

CHAPTER IV.

SECTION I.—GENERAL PRINCIPLE OF RATIONAL PRACTICE—DICTATED BY THE PATHOLOGY OF THE DISEASE—CONFIRMED BY OBSERVATION AND EXPERIENCE.

IT has been observed that the essential characteristic, the leading and most prominent indications requiring special attention and permanent relief, are the "Algide," or loss of temperature; the loss of nerve-power in the ganglionic and pneumogastric nerves and their branches; the altered or disorganized condition of the blood; the impaired or obstructed circulation, and the early and direct tendency to congestion; and that these prominent and essential features are correspondingly developed, and in their relation to each other are too intimate and dependent to admit the idea of priority and regular order of succession. The primary impression being on the blood, these proceeding, *pari passu*, together constitute the complex character of the disease; and suggest the general principle of rational practice. If our pathology be correct, it must be regarded as the foundation and only base for a successful mode of treatment, and must be allowed to dictate the maxims of rational practice in the prevention and cure of this singular disease. The neglect to apply to the treatment of the cholera the science of its peculiar and established pathology and phenomena, or to give heed to its teachings, has no doubt led to the errors and failures in practice, which, from their too general occurrence, induced the learned and celebrated Dr. Velpeau to declare, before the Academy of Medicine in Paris, that " we know nothing more of the treatment of cholera now, than on its first appearance in 1832. All our remedies and modes of practice have failed."

By observing the fundamental principles

of the science of medicine, and adopting a mode of practice suggested by the pathology and phenomena of the cholera, these errors and failures, which have justly brought odium upon the so-called regular profession, will probably result in saving nine-tenths of those attacked, instead of losing that appalling proportion, as has been the case in some instances in years past.

What, then, is the principle which, for nearly half a century, has been strangely overlooked, and utterly disregarded by the so-called regular profession, so far as the maxims of rational practice are concerned in the treatment of this disease? We unhesitatingly affirm the principle suggested by the pathology of the disease is, and must be, one that will reproduce and re-supply the lost caloric, or restore warmth to the body ; one that will restore promptly the lost nerve-power to the ganglionic nerves especially ; one that will arrest and remove the tendency to congestion, equalize the circulation and relieve the oppressed respiration, and thus mitigate the long train of dependent symptoms. For this purpose, a prompt and diffusive stimulant is required of sufficient power to meet these urgent demands, and suspend promptly any further depressing influence or action of the cholera poison. A stimulant, essentially different from alcohol in any of its forms, is required. Alcohol, except so far as it necessarily enters into the composition of medicines, is inadmissible. So, too, are all those stimulants whose action is violent, or tends to induce constitutional derangement, or impairs in any way the subsequent health of the patient. It must be one prompt, kind and diffusive in its nature, and peculiarly adapted to meet and relieve the essential urgent symptoms on which the whole train of *non-essential symptoms* depend. In short, it must be one possessing the singular properties of a stimulant, sedative and astringent, especially an arterial stimulant and antispasmodic.

In confirmation of this doctrine, we may refer to the general principle exhibited in the most successful modes of practice. During the prevalence of the cholora in 1832, the physician having charge of the Tenth Ward in this city, in which more than a thousand cases occurred, adopted as the principle of general practice in that ward a prompt and diffusive stimulant, which was, at that early day, regarded by him as based on the pathology of the disease. This principle was strictly observed and fully carried out in practice by all his assistants. The result, embracing the different stages of the disease, and some of the most malignant cases, was the curing and saving of more than nine-tenths of those attacked.

Another instance directly in point is the course pursued by Mr. Forward, an unprofessional gentleman, who had over two hundred laborers in his employ, among whom the cholera prevailed in 1832 with its accustomed severity. On its first appearance, Mr. Forward, unadvised, and depending on ordinary domestic remedies, adopted as the base of practice in the emergency a prompt and diffusive stimulant, which proved perfectly successful. Being advised to continue the same course, should any more cases occur, the result was, in treating a large number of cases, including thirteen in his own family, that all were cured. Again, on the reappearance of the cholera in 1835, the same practice was pursued, with the same uniform success. Can anything be more satisfactory or more conclusive as to the adaptation of a principle of practice to the pathology of the disease, or furnish better evidence of the correctness of the doctrine we have advanced?

Richard Brown, Esq., surgeon, Cobham, Surrey, November, 1848, reports a case treated successfully by rectified oil of turpentine, the therapeutic character of which is unquestionable.

Dr. Massie, of Texas, adopted a similar principle of practice, and highly commended the same to his professional brethren, as the safest, best, and most efficient in the treatment of the cholera. He affirms, that of all the modes devised for the prevention and cure of this disease, none is so simple and efficacious as the one exhibited in his practice.

The homœopathic treatment, which claims to be a complete and perfect system, arranged and adopted by its originator and all his disciples, confirms the correctness of the doctrine we have advanced. Its curative principle in the treatment of cholera is based on a prompt and diffusive stimulant, peculiarly adapted, so far as it has any power, to meet and relieve the essential symptoms of this disease. Hence its success and favorable results, which show a saving of nearly nine-tenths of all the cases treated.

Again, the eclectic physicians, who now, including all of the reform school, constitute a majority of the practitioners of medicine in this country, adopted a principle essentially similar, which has governed their practice in the treatment of this disease from its first appearance in 1832. Their system seems to have been more strictly conformed to the pathology of the cholera than that of any other school. Hence, their unparalleled success furnishes the most substantial and conclusive evidence, sustaining the correctness of the doctrine we have adduced, and the general principle of rational practice suggested and imperatively demanded, by the pathology of the disease. Their treatment, directed mainly to the relief of the essential symptoms, has been based on a prompt and diffusive stimulant, which, fulfilling to some extent the indications required, has enabled them to meet the disease on each occasion of its reappearance with some assurance of success, and more generally to arrest its progress or subdue its power as exhibited in its several stages, and even in many instances to restore the patient and save life in the last stage of the almost hopeless collapse. This is clearly shown in the actual results which fully exhibit the incomparable fact that in private practice considerably more than nine-tenths of the cases of "true cholera" are cured, and the constitution and health of their patients saved unimpaired.

Again, this doctrine is substantially confirmed by the results of the experiments made by Drs. Hill and Davies, in the exhibition of chloroform, either alone or combined with other stimulants. In the carefully detailed account of its exhibition in the various stages of the disease, it is clearly shown that its direct action tends to arrest and suspend the depressing influence of the primary cause, and when properly combined with other stimulants, affords very prompt relief. The favorable results thus obtained encourage the hope that it may prove a successful remedy and lead to the adoption of a more consistent mode of practice in the treatment of epidemic cholera. In India, in Europe, and in America, it is now regarded as a very important remedy, and especially indicated in this disease. As an antidote to miasmatic poison, and as a prompt and diffusive stimulant when properly combined, it is admirably calculated to meet and suspend the most urgent symptoms. In short, it may be considered, in relation to this disease, an excellent therapeutic agent, and well calculated to form the base of the principle for which we contend.

But again, our doctrine is confirmed by the experiment usually termed "venous transfusion." The solution of soda, when raised to a temperature from 105° to 120° Fahr., and injected into the veins of the suffering patient, gave *temporarily* prompt and immediate relief; but, when injected at a lower temperature, failed. In this experiment, the sole and only agent contributing to the result was, as before explained, the free caloric which immediately permeated every tissue, supplied warmth to the body, relieved the depressed nerve-power, equalized the circulation, and restored generally the normal action of the system. Of this result, and of the diffusive and prompt stimulating power of free caloric, there can be no question. The principle here evolved, which answered so perfectly the imperious demand and so immediately suspended the power of the disease, is the very principle dictated by its pathology. Stronger and better evidence of the utility of a prompt and diffusive stimulant, permanent in its character and influence, cannot be furnished ; one that will act kindly, without violence and without any disturbance to any organ or tissue, to injure or delay the return of immediate and perfect health after the disease is subdued. Such we affirm to be the principle demanded in the successful treatment of the epidemic cholera.

Section II.—Remedies, Recipes, etc.

Considering the general principle of treatment, and the nature of the remedy so clearly suggested by the pathology of the disease to be fully established, it now remains for us to point out some of those curative agents which may be employed to advantage. It may be here observed, that among the few that can be confidently recommended, there is no single remedy yet discovered which seems to possess all the properties necessary to meet the complex condition presented in a malignant case of cholera. Yet it is believed we have simple remedies, which, when properly combined, will prove successful. Among the number that seem best adapted to meet and fulfill the indications, may be named chloroform, as the leading remedy on which we may reasonably hope for success. This may be united with spirits of camphor, the tincture of xanthoxyli fraxinifolii bacca, the compound fluid extract of rhubarb and potassa,* and the oil of monarda punctata, and a very valuable and reliable remedy obtained. The following formula exhibits the mode of combination, which may be varied and adapted to suit any emergency :

℞ Chloroform, (sq.,) . . . . . ʒij.
   Spirits Camph., ... ʒj.
   Ol. Monarda, ... gtts. x.
   M. et adde—
   Tinc. Xanthox. Frax. Bac., ℨij.
   Fluid Ext. Rhei et Potas., ℨiv.

M. S.—From ʒj. to ℨss. every half-hour, hour or two hours, according to the urgency of the symptoms and the stage of the disease. As soon as relief is obtained, it

* See American Dispensatory.

should be given in minimum doses and less frequently. This is admirably adapted to the cold stage, and will give prompt relief in a great majority of cases.

In the premonitory stage, it can be administered to good advantage in small and less frequent doses. In some instances, an additional astringent may be necessary. The deceptive and painless diarrhœa should receive prompt attention, and be regarded and treated as the incipient form of the disease. According to the best authorities, the diarrhœa commences with the first chemical change or alteration of the blood, and proceeds gradually, in most cases, for some hours, and even in some instances, though rarely, for days. It is not sufficient to check the diarrhœa merely; the cause must be removed, which is essentially of miasmatic origin. When the cholera is prevailing, and the diarrhœa is essentially choleraic, or the result of a depressing miasmatic influence, it should be treated with chloroform, aided, if required, by appropriate astringents.

In the fully developed stage, and even in the stage of collapse, perhaps no combination is better adapted to meet promptly all the necessities and wants of the system, and suspend the action of the cholera-poison, than the one named above. It is a simple, prompt and diffusive stimulant, approximating the principle indicated. This peculiar remedy is essentially required, and should be continued through all the stages of the disease till relief be obtained, varying its administration according to the urgency of the symptoms. When the stomach is too irritable to retain medicine, it should be given by the bowel. Take of the above mixture, one-half ounce, of the tincture of prickly-ash berries one-half ounce, of the tincture of opium ten drops, of warm water one ounce and a half—mix and inject. This may be repeated after every evacuation three or four times, unless relief be obtained earlier. Thus, it should be administered perseveringly by stomach and by bowel, aided by due employment of all necessary external means for furnishing warmth and giving relief. Opium, however, should be omitted after two or three injections. Its continued use to check the movement of the bowels is decidedly injurious.

The vomiting and irritability of the stomach may often be allayed by a strong decoction of spearmint and horse-peppermint (monarda punctata), equal parts, alternated with camphor water in small repeated doses every five minutes. This will often succeed when all other means fail.

The compound cajeput mixture* is a very excellent and prompt stimulant, and may be alternated with other remedies with good effect. It is particularly useful in allaying violent cramps, and restoring warmth to the body, and may be given in doses of one teaspoonful every ten or twenty minutes in mucilage, simple syrup, or, better still, in hot brandy-and-water sweetened.

The aromatic tincture of guaiac* will be found very useful in some cases, and may be united with chloroform according to the following :

* See American Dispensatory.

℞. Chloroform, (sq.) . . ℨij.
Spirits Camphor, . . ʒj.
Ol. Monarda, . . . gts. v.
M. et adde—
Tinc. Guaiac. Arom., ℨiv.
M.

S.—From one-half to one tea-spoonful every half hour, or, if necessary, in violent cases every twenty minutes, in a little sweetened water. This may be alternated with some other remedy to great advantage.

Chloric ether has been with some a very favorite remedy, and, in combination with other diffusive stimulants, may serve a good purpose. So, too, the spirits of turpentine, and the rectified oil of turpentine, have proved very beneficial, the former in combination, the latter administered alone. These agents, however, can be rendered more prompt and effective by combination. It is the promptness, the instantaneous or electric action like that of oxygen, ozone, and caloric that gives value to the combination, and renders it peculiarly efficacious when it possesses the other peculiar properties required.

In the early stage, sulphuric acid, in the form of elixir vitriol, has given very prompt relief, and is very highly recommended as a curative agent in the treatment of this disease. The following formula presents the mode of its exhibition:

℞. Elixir Vitriol, ... ʒj.
    Tinc. Xanthox. Frax. Bac. ʒij.
    Ess. Lemon, ... ʒj.

M.— S.—Tea-spoonful in a gill of sweetened cold water every two or three hours.

This recipe was used in the incipient stage quite extensively in the epidemic of 1849, with decided advantage. It generally removed the symptoms speedily, without any other treatment. In the more advanced stage it was thought not so reliable as other means named above.

Dr. Fuller, of this city, advocates the use of sulphuric acid as a prompt and efficient remedy, and affirms that according to his experience, a great majority of cases may be cured by this mode of treatment.

Dr. Cox, of England, has also spoken in its favor, and recommended its use as an infallible remedy. The eclectic physicians are entitled to the credit of its first introduction as a curative agent in the treatment of the Asiatic cholera, combined with the tincture of prickly-ash berries and the essence of lemon, as noticed above. In our estimation it may be rendered more effective, combined according to the following:

℞. Elixir Vitriol, } .. āā., ʒj.
    Chloric Ether, }
    Tinc. Xanthox. Frax. Bac. ʒij.
    Ess. Lemon, ... ʒj.
M.

S.—A tea-spoonful in a gill of sweetened cold water every two or three hours. Thus combined, it forms a very prompt and diffusive stimulant, and is well adapted to meet the indications in the earlier stage of the disease. In the last stage perhaps no remedy will be found so prompt and decided in its action as the injection named above, with the internal use of chloroform as combined in the recipe on page 189.

In cases of excessive irritability of the stomach, the following combination was administered with good effect, and was especially beneficial in cases attended with stupor from the commencement of the disease:

℞. Common Salt, ... ʒj.
Black Pepper, ... ʒj.
Vinegar, ... f. ʒv.
Hot Water, . . . f. ℥iv.
M.

Of this, when settled, or strained, a table-spoonful may be given every ten or twenty minutes. It seldom failed to quiet the stomach and check the motion of the bowels. In this condition the injection should be also administered, and repeated as occasion may require.

Some advocate the use of the spirits of ammonia and tincture of capsicum, properly combined with other diffusive stimulants, as a very efficient and successful remedy. The following is, perhaps, the most desirable formula:

℞. Chloroform, (sq.)
　　Spts. Camph.,
　　Spts. Ammonia Aromat.,　} āā., ℨ iij.
　　Tinc. Capsicum,
　　Elix. Opii (McMunn's), . . ℨ ss.
　　Syr. Zingiberis, ... ℥ ij.

M.—S.—Tea-spoonful in water every thirty minutes till relieved. Then less frequently, according to circumstances. This is said to give very prompt relief in the earlier stage of the disease. With some practitioners the following has been quite a favorite remedy :

℞. Æther Chloric., . . . - ℥ j.
　　Tinc. Cardamom., . . . ℥ ij.
　　Spts. Camph., ... ℥ ss.
　　Elix. Opii (McMunn's), . ℨ ss.
　　Syr. Zingib., ... ℥ ij.
M.

S.—Two tea-spoonsful in water every 10 or 30 minutes till relieved, then continued less frequently and in less doses every one, two, three, or four hours, according to circumstances.

For the purpose of promoting reaction in cholera and diarrhœa, the following formula has been extensively used and most universally approved. It is, indeed, so highly valued in England and in India, that it is ordered to be always in store and in readiness in the Medical Field Companion of the army when on the march:

℞. Ol. Anisi,
    Ol. Cajeput, } . . . āā., 3 ss.
    Ol. Juniper,
    Æther Chloric, ... ℥ ss.
    Liquor Acid. Haleri,* . . 3 ss.
    Tinc. Cinnamon, . . . ℥ ij.
M.

S.—Ten drops every fifteen minutes, in a table-spoonful of water. An opiate may be given with the first and second dose, but should not be continued.

Another recipe which has been used with some success in private practice, illustrative of the use of chloroform as a diffusive stimulant and sedative, is the following:

---

* Sulphuric acid, one part; rectified spirit, three parts.

℞. Chloroform (sq.) ⎫
   Spts. Camph., ⎪
   Tinc. Capsicum, ⎬ . āā., ʒij.
   Tinc. Zingib., ⎪
   Tinc. Cardamom., ⎭
   Syr. Simplex, ... ℥ij.
M.

S. Tea-spoonful in a little water every half hour, hour, or two hours, according to circumstances. An opiate may be given with the first and second dose, but should not be continued. Should the first dose be ejected, give another immediately after the vomiting.

In collapse, which is simply a more advanced stage of the disease, indicating the gradual failing of all the powers of life, our main reliance is on enemata, as noticed above, often repeated, and continued as occasion may require.

Rev. Dr. Hamlin, of Constantinople, observes, "It is difficult to say when a cure has become hopeless. The blue color, the cold extremities, the deeply sunken eye, the vanishing pulse, are no signs that the case is hopeless. Scores of such cases in the recent epidemic have recovered."

Here it may be proper to add, that a cure, even with the most efficient remedies, cannot be easily effected without placing the patient at the commencement in a recumbent position. This appears indispensable. The patient should be placed in bed and kept there in the horizontal position, comfortably covered with blankets, and with warm applications to the feet. Every necessary convenience should be at once provided to prevent, if possible, the patient from rising to, or standing upon, his feet, for the erect posture, before relief is fully obtained, will inevitably hasten the unfavorable termination of the disease. On this direction, therefore, the physician must insist if he would save his patient. Says an eminent physician, perfectly familiar with the disease, "This direction faithfully observed, and good nursing, will save very many patients even without medicine."

Of the auxiliary aids, consisting of various external applications, we cannot speak in very flattering terms. To the mind of the practitioner the more important are readily suggested, and are promptly employed by nurses in the earlier stages of the disease. It is impossible for any person to attend on a case of true cholera without being instinctively moved to apply heat friction, and warm stimulants to the surface for the relief of the suffering patient. Any attempt to prevent these kind offices and apparently beneficial appliances would be unwise, and most certainly, in private practice, unavailing. It becomes, therefore, necessary to direct the use of those which are most agreeable to the patient and tend to preserve and sustain the recuperative power; those which tend to weaken and depress the system are the most objectionable. Among the number that seem to do good, we may mention bottles of hot water to the feet and calves of the legs, hot bricks dipped in water and wrapped in flannel and applied to different parts of the body; blankets wet in water as hot as can be borne, and wrung out so as not to drip, and applied to the whole surface, and changed at short intervals, so as to keep up a steady and permanent temperature of the surface; flannels moistened with spirits of turpentine, or other stimulant embrocation, and laid over the stomach and bowels, may be employed, as these all, in some instances, seemed to be beneficial. Their necessity and use, however, must be governed by circumstances. As we have before said, our main reliance is on a prompt and diffusive stimulant internally; other means, at best, are very uncertain.

Such are some of the remedies evidently suggested by the pathology and phenomena of the disease, and adapted to meet and remove the more urgent, essential symptoms. They are not entirely new. They have been employed to some extent in former epidemics of cholera, and have sustained a good reputation as useful and curative agents in the treatment of this disease. The combinations here suggested are the result of observation and experience, and are intended to present the form in which these remedies can be exhibited to the best advantage. They are simple, prompt, and reliable, such as will leave the system, when the disease is subdued, in its ordinary condition, without any injury whatever to prevent its immediate return to its normal state of health. Let them be employed, and their utility thoroughly tested. They will bear the strictest scrutiny, and sustain their reputation untarnished under the most trying circumstances. Should the cholera appear again in our midst in its epidemic form, and these remedies be generally employed and properly administered, we venture to predict their efficacy will be abundantly proved in the successful result of saving more than nine-tenths of those attacked.

SECTION III.—PROPHYLAXIS—OR MEANS OF PREVENTION.

IN presenting a course of preventive treatment consistent with the origin and general character of the disease, we are necessarily limited to the means of sustaining the *normal* action of the system, and suppressing the operation of those causes which, by reducing the general health, tend to generate, foster, and develop the cholera. Of the former so much has been written and published, inculcating the general principles of hygiene, that it seems quite unnecessary to dwell on a subject so familiar to the great mass of community; yet, there are occasions when the most familiar truths have to be impressed upon the mind, by constant repetition, to prevent threatened dangers, and obviate the most serious consequences. In no instance is this more important than in time of prevailing epidemics; for it is an undeniable fact, that multitudes *will* neglect the most obvious principles of hygiene, and tolerate, with utter indifference, the most offensive nuisance, in and around their dwellings, and if attacked by disease, will often wonder why *they*, more than *others*, should be visited by a malignant disease, or become the victims of a prevailing epidemic. Hence the necessity of urging the observance of some of the most obvious principles of hygiene, in the preventive treatment of Asiatic cholera.

Pure air, pure water, and a frugal nutritious diet are Nature's great preventives for the thousand ills of life. These are the great essentials in sustaining the healthful and normal condition of the system, always of primary importance in preserving its tone and energy, and rendering it impervious to any miasmatic or epidemic influences. Therefore, the tone of the system should, more especially when epidemics are prevailing, be kept fully up to its normal standard. This cannot be accomplished without pure air,—whether our dwellings be located in the city or in the country; free ventilation of all apartments is of the first importance. Kitchens, sitting-rooms, dressing-rooms, and especially sleeping-rooms, should be kept constantly and thoroughly ventilated; cellars and vaults, too, should receive attention, and be kept free from a deteriorated or foul atmosphere. Everything within and without our dwellings, tending to impregnate the atmosphere with noxious effluvia, should be removed, and the foul air promptly purified by the use of appropriate disinfectants.

Pure water for drinking and culinary purposes is another preventive remedy, whose employment cannot be safely omitted. It is a well-known fact that, in various localities, wells only a few feet deep, which are mainly supplied by drainage or surface water, have proved a fruitful source, and in some instances a direct and efficient cause of epidemic cholera.

The water from rivers flowing past large cities and villages is often so impure as to render its use decidedly deleterious, if not an actual source of disease. In some cases they have been literally so filled with portions of fish, and other animal matter, that all city supplies were made endurable only by long-continued filtration. The waters of many of our Southern and Western rivers are rendered impure from the lime and surface drainage with which they are so highly impregnated that they often become a direct source of diarrhœa and cholera. Pure water, free from the impregnation of vegetable, animal and mineral substances, should be sought and obtained for domestic use.

A good nutritious diet is an indispensable requisite in the prevention of disease. The system in comparative health requires, and should regularly receive, its proper aliment. Its daily recurring demands should be judiciously met with pure and wholesome food, in such quantity as can be readily digested, assimilated and duly appropriated for the supply of its wants. Due regard, however, must be had to the existing and peculiar condition of the digestive organs, on which mainly depends the process of supporting and perpetuating the general health.

It is not the profuse variety and the incongruous mass composed of baked, roasted, boiled and fried meats, fish and fowl, oyster, lobster, frog and turtle, with puddings, tarts, jellies, cakes and creams from the pastry room—fruits and salads, native and foreign, rich and rare—alcoholic stimulants, and cooling ices, but the simple, plain and frugal diet, properly cooked and particularly nutritious, that conduces to the most vigorous health.

Regular, temperate habits in all things, are especially commended; excesses of all kinds are reprehensible. Great and sudden changes in the habits of living are always deleterious, and must be particularly so, when an appalling and fatal epidemic is prevailing. Temperance, sobriety and cheerfulness, regular hours for meals, for rest and for business, repeated ablutions and perfect cleanliness, moderate exercise and avoidance of irregularities, persevering self-government and duly subjected passions, all contribute to health, to happiness, and the prevention of disease.

Exposure to the extremes of heat and cold should be avoided, and the clothing properly adapted to the climate—to the season and its variable temperature. Constant vigilance is necessary to guard against the numberless causes tending to produce an abnormal condition, resulting in the derangement of the stomach and bowels, or in depressing the nervous power, thus enfeebling and prostrating the general health. The neglect of these hygienic principles and essential preventives of cholera may induce the condition which temptingly invites the disease. Some are vastly more susceptible than others, and may not be able, with all their watchfulness and care, to avoid an attack, should the disease extensively prevail among us.

The premonitory symptoms requiring special attention, when the epidemic cholera is prevailing, are definitely presented in Chap. II., Sec. 2, page 56, to which special reference is made. Whenever any of these do occur, though generally supposed to present no particular characteristic of the cholera, they should, however, receive prompt attention. The loss of animation, the depression of nerve-power, the pain in the forehead and slight vertigo, the nervous agitation and oppression at the chest, with slight nausea, may in most instances be promptly removed. They should be at once patiently and perseveringly treated by the use of camphor water, prepared as follows : Take spirits of camphor, one tea-spoonful, and put it into a half-pint of cold water, and give of the mixture two tea-spoonfuls every half-hour, hour, or two hours, according to the severity of the symptoms. A strong decoction, or tea of horsemint (monarda punctata), is an excellent remedy even in this early stage. The essence of monarda, or horsemint, in doses of eight or ten drops in a little water, and repeated every hour or two, will often give prompt relief. Where the horsemint cannot be obtained, the spearmint, and the peppermint also, may prove serviceable.

Keith's concentrated Tincture of Veratrum Viride is also an excellent remedy in these premonitory symptoms. Put three or four drops into a tumblerful of cold water, and give of the mixture a tea-spoonful every hour or two hours, as occasion may require. This may be alternated with the essence, or tea of horsemint.

But another more general symptom, which may be properly termed the incipient

PROPHYLAXIS.

stage of the disease, is the slight diarrhœa, usually termed painless, though it is by no means always so, but frequently the very reverse, severe and painful. This at first may be slight, but gradually increasing, soon becomes obstinate, painful, and exceedingly difficult to control. It therefore should receive attention at its very commencement, for it is in reality the stealthy invasion of the citadel—it is the cholera. The loss of life becomes imminent; treatment becomes indispensable; send at once for your physician. And, in the meantime, continue the camphor mixture, the horsemint tea, and give of the fluid extract of rhubarb and potassa, prepared according to the formula in the American Dispensatory, one or two tea-spoonfuls every hour, and, if necessary, add four or five drops of laudanum, or its equivalent in paregoric, to each dose, till relieved. In this early stage, opium in small doses may be given, four or five times, but should not be continued. These remedies, properly administered, will control the great majority of cases.

If, however, the diarrhœa be uncontrolled and vomiting ensue, the recipe on page 189 will be found very efficient, and should be perseveringly administered till relief is obtained. It is prepared as follows : Chloroform, two drachms ; spirits of camphor, one drachm ; essence of monarda (or horsemint), three drachms ; tincture of prickly-ash berries, two ounces ; fluid extract of rhubarb and potassa, four ounces—mix. Give from one-half to one table-spoonful every half-hour, hour, or two hours, according to the urgency of the symptoms and the stage of the disease. This remedy is well adapted to every stage, and may be used in collapse as an injection, combined as follows : Take of the above mixture *two table-spoonfuls*, and add to it tincture of prickly-ash berries, *two table-spoonfuls;* laudanum *ten drops;* warm water, *six table-spoonfuls*—mix, *and inject up the bowel*. This injection should be repeated as often as required. In some desperate cases it has been repeated many times and the patients saved.

Wherever the disease prevails, all discharges from cholera patients should be promptly disinfected and disposed of. Bedding, linen, water-closets, cesspools, etc., should be thoroughly disinfected and renovated, so that no germ may remain to propagate the disease.

FORMULÆ

FOR SOME OF THE PREPARATIONS USED IN THE ABOVE RECIPES.

GREENHOW'S AROMATIC TINCTURE OF GUAIACUM.—Take of guaiacum, cloves and cinnamon, each, in powder, *one ounce;* best brandy, *two pints.* Macerate for fourteen days and filter.

Dose.—From a tea-spoonful to a table-spoonful, in sweetened water, every fifteen or twenty minutes. —*Am. Dis.*

COMPOUND CAJEPUT MIXTURE—HUNN'S DROPS.—Take of oils of cajeput, cloves, peppermint, and anise, each, *one fluid ounce;* rectified alcohol, *four ounces.* Dissolve the oils in the alcohol.

The ordinary dose is from ten drops to half a tea-spoonful; to be given in simple syrup, mucilage of slippery-elm, or in hot brandy and water *sweetened.* —*Am. Dis.*

FLUID EXTRACT OF RHUBARB AND POTASSA.—
Take of the root of the best India rhubarb, in powder, and bicarbonate of potassa, of each, *one ounce;* cassia or cinnamon, and golden seal, in powder, of each, *half an ounce;* boiling water, one-half pint. Macerate the roots and seeds for an hour; strain and dissolve the potassa in the strained liquor when nearly cold, and add one gill best brandy; essence of peppermint, one tea-spoonful, and refined sugar, *two ounces.*

Dose.—From one to two tea-spoonfuls as often as necessary.—*Am. Dis.*

TINC. XANTHOXYLI, or Tincture of Prickly-ash Berries.—Take of prickly-ash berries *eight ounces;* diluted alcohol, *two pints.* Form into a tincture by maceration, or displacement, and make two pints of tincture.

The ordinary dose is twenty or thirty drops. In cholera, from a tea-spoonful to one or two table-spoonfuls, according to circumstances.—*Am. Dis.*

TINCTURE OF OIL OF MONARDA—Essence of Monarda, or Horsemint.—Take of oil of horsemint *one fluid ounce;* alcohol, *nine fluid ounces,* Imp. Meas. Mix with agitation.

Dose.—From ten to twenty drops on sugar, or in sweetened water.—*Am. Dis.*

ELIXIR OF OPIUM, prepared on the base of Dupuy's formula is less objectionable as an ingredient in recipes for an advanced stage of cholera than other preparations of that drug.

www.ingramcontent.com/pod-product-compliance
Lightning Source LLC
Chambersburg PA
CBHW020815230426
43666CB00007B/1021